5th International Conference on Architecture, Materials and Construction (ICAMC 2019)

IOP Conference Series: Materials Science and Engineering
Volume 809

Lisbon, Portugal
2 - 4 December 2019

ISBN: 978-1-7138-2801-3
ISSN: 1757-8981

Printed from e-media with permission by:

Curran Associates, Inc.
57 Morehouse Lane
Red Hook, NY 12571

proceedings.com

Some format issues inherent in the e-media version may also appear in this print version.

Institute of Physics
Dirac House, Temple Back
Bristol BS1 6BE UK

Phone: 44 1 17 929 7481
Fax: 44 1 17 920 0979

techtracking@iop.org

Additional copies of this publication are available from:

Curran Associates, Inc.
57 Morehouse Lane
Red Hook, NY 12571 USA
Phone: 845-758-0400
Fax: 845-758-2633
Email: curran@proceedings.com
Web: www.proceedings.com

TABLE OF CONTENTS

IOP Conf. Series: Materials Science and Engineering **809** (2020) 011001 doi:10.1088/1757-899X/809/1/011001

Preface

The present document exposes the significant contribution to scientific knowledge gathered in the 5th International Conference on Architecture, Materials, and Construction (ICAMC 2019) and its workshop, which is the 3rd International Conference on Nanomaterials and Biomaterials (ICNB 2019). Both held at the *Faculdade de Arquitectura* of the *Universidade de Lisboa*, Portugal, during December 2sd to 4th, 2019, in partnership with the Hong Kong Chemical, Biological & Environmental Engineering Society (HKCEEBS).

The conferences provided a broad platform for participants' communications on their research outcomes. And also, to the keynote speakers and their meaningful speeches on their work's progress on nowadays topics, which should wake the consciences to the coming future, all under a fulfilling academic atmosphere, and rewarding cultural experience.

On behalf of the organizing committee, it is a great honor to announce the conference proceedings, which gathers a selection of submitted papers related to Civil and Structural Engineering, Architecture, and Urban Planning, Materials Manufacturing, and Processing, Nanomaterials, Biomaterials, from prime worldwide universities, research institutes, and industries.

Lastly, I would like to emphasize the work and contributions of all my colleagues: the authors by quality of their papers; the reviewers by the unparalleled dedication to the reviewing process and the added contributions; the organization professionals by commitment to the conference accomplishments; and the proceedings team by the determination to enable this remarkable volume.

Professor Nuno Dinis Cortiços
Faculdade de Arquitectura of the *Universidade de Lisboa*, Portugal
Lisbon, February 13th, 2020

IOP Conf. Series: Materials Science and Engineering **809** (2020) 011002 doi:10.1088/1757-899X/809/1/011002

Conference General Co-Chairs

Prof. Nuno Dinis Cortiços, Universidade de Lisboa, Portugal
Prof. Eric J. Strauss, Michigan State University, USA
Prof. Marco Casini, SAPIENZA University of Rome, Italy
Prof. Jordi Arbiol, ICREA & Catalan Institute of Nanoscience and Nanotechnology (ICN2), CSIC and BIST
Prof. Xiang Zhang, University of Cambridge, UK

Program Co-Chairs

Prof. Nilson Tadeu Mascia, University of Campinas, Brazil
Prof. Kwang Leong Choy, University College London, UK
Prof. Luc Avérous, University of Strasbourg, France

Technical Committees

Prof. Ramzi Taha, Qatar University, Qatar
Prof. David Chua Kim Huat, National University of Singapore, Singapore
Prof. Ruzica Bozovic Stamenovic, National University of Singapore, Singapore
Prof. Wassim E. Raphael, Department of Civil & Environmental Engineering, Lebanon
Prof. Antonio Zuorro, University of Rome, Italy
Prof. Kazuo Umemura, Tokyo University of Science, Japan
Prof. Li Shang, Northwestern Polytechnical University, China
Prof. Arief BUDIMAN, Singapore University of Technology & Design, Singapore
Prof. Indriana Kartini, Universitas Gadjah Mada, Indonesia
Prof. K. BYRAPPA, Mangalore University, India
Prof. O.L. Shanmugasundaram, K.S. Rangasamy College of Technology, India
Prof. Djezzar Mahfoud, University Mentouri Bros Constantine Algeria
Prof. Khalid B. Najim, University of Anbar, Iraq
Prof. Toufik Boufend, Constantine 1 University Algeria
Prof. Mohammad Arif Kamal, Aligarh Muslim University, India
Prof. D. Edison Selvaraj, Panimalar ENGG College, India
Prof. Taiki Saito, Toyohashi University of Technology, Japan
Assoc. Prof. Stephen Belton, University of Florida, USA
Assoc. Prof. Usama Ebead, Qatar University, Qatar
Assoc. Prof. Kenneth Loh, University of California, San Diego, USA
Assoc. Prof. Wael Zaki, Khalifa University, Abu Dhabi, UAE
Assoc. Prof. Fei Xu, Northwestern Polytechnical University, China
Assoc. Prof. Keyu Xie, Northwestern Polytechnical University, China
Assoc. Prof. Hua LI, Nanyang Technological University, Singapore
Dr. Hamed Bolandi, Michigan State University, USA
Dr. Silvia Mazzetto, IUAV University of Architecture in Venice, Italy
Dr. Ana Almerich-Chulia, Universitat Politecnica de Valencia, Spain
Dr Lau Siu Kit, Eddie, National University of Singapore, singpore
Dr. Djamel Boussaa, Qatar University, Qatar

IOP Conf. Series: Materials Science and Engineering **809** (2020) 011002 doi:10.1088/1757-899X/809/1/011002

IOP Conf. Series: Materials Science and Engineering **809** (2020) 011003 doi:10.1088/1757-899X/809/1/011003

First Name	Last Name	Institution	Email Address	
Jie	Liu	China Institute of Water	lj_iwhr@163.com	B19-
Raffaella	Maddaluno	Faculty of Architecture	rmaddaluno@fa.ulisb	B19-2009
Kaixin	Hu	Chongqing Jiaotong	huxingyishay@163.co	B19-2012
Yanming	Wu	Southeast University,	791367448@qq.com	B19-
Xiaoping	Zhang	Wuhan Harbor Engineering	shoppinggre@hotmai	B19-2013
Katarzyna	Fiedoruk	Bialystok University of	k.gladyszewska@pb.e	B19-2014
Hechi	Wang	Hubei University of	wanghechi@163.com	B19-2015
Amr Mohamed	Mostafa	The British University in	amr.ibrahim@bue.ed	B19-
Dorota	Krawczyk	Bialystok University of	d.krawczyk@pb.edu.p	B19-2018
Taiji	Mazda	Kyushu University , Japan	mazda@doc.kyushu-	B19-2021
Wegdan	El-Nadoury	Pharos University in	Wegdan.wagdy@pua.	B19-2022
Sahar Ali	Ismail	Université Saint-Joseph de	Sahar.ismail@net.usj.	B19-
Anna	Stefanska	Warsaw University of	Anna.Stefanska2@pw	B19-
Joanna	Pietrzak	Warsaw University of	joanna.pietrzak@pw.	B19-
Syed Ishtiaq	Ahmad	Bangladesh University of	siahmad@ce.buet.ac.	B19-2026
Sahar Ali	Ismail	Université Saint-Joseph de	Sahar.ismail@net.usj.	B19-
Hikmat	Daou	Saint-Joseph University ,	fi.esib@usj.edu.lb	B19-2031
Antonio Rodero	Serrano	Universidad de	a.rodero@uco.es	B19-
Elise Hanna	ZGHEIB	Saint-Joseph University ,	elise-	B19-
Adelya	Khayrova	Research Centre of	adelya15@mail.ru	B19-1001
Zhuan	Zhao	University of Padova, Italy	zhaozhuan2008@gm	B19-
Jose Angel Roldan	Lopez	Universidad Privada	jroldanl@yahoo.es	B19-1007
Lujie	Peng	University of Glasgow, UK	lujiepeng219@qq.co	B19-
Lincoln Almeida	Cavalcante	Instituto Federal Do	l.a.cavalcante@outlo	B19-

Peer review statement

All papers published in this volume of *IOP Conference Series: Materials Science and Engineering* have been peer reviewed through processes administered by the proceedings Editors. Reviews were conducted by expert referees to the professional and scientific standards expected of a proceedings journal published by IOP Publishing.

IOP Conf. Series: Materials Science and Engineering **809** (2020) 012001 doi:10.1088/1757-899X/809/1/012001

Pier Luigi Nervi and the School of Italian Engineering: *Science or the Art of Building*?

MADDALUNO Raffaella

Lisbon School of Architecture, CIAUD, Portugal

rmaddaluno@fa.ulisboa.pt

Abstract. The recent and painful event of the collapse of the Riccardo Morandi bridge in Genoa has brought to light in international debates what, in the second half of the 1900s, was recognized as the School of Italian Engineering. The sixties mark a pinnacle in the relationship between technical and scientific studies, the evolution of infrastructural networks, and a specific attention to the importance of form in the final result of engineering works. The Stadio Flaminio, designed by Pier Luigi Nervi (1891-1979) on the occasion of the 1960 Olympics, internationalized an engineering approach that can be identified as *made in Italy*. This paper aims to analyse, starting from a charismatic figure like Nervi, what the characters of this identity are and if, in light of recent failures, we can talk about a school.

1. Introduction

Around Nervi, personalities of the caliber of Morandi, Sergio Musmeci intertwine, who contribute, through sharing or through contrasting positions, to interpreting this heritage. The realizations of the bridge over the Arno to Incisa by Silvano Zorzi (1962) and the cable-stayed bridges by Riccardo Morandi on the Maracaibo lagoon in Venezuela (1962) and on the Polcevera in Genoa (1964), mark the highest moment of this art of building and of the Italian research on prestressed reinforced concrete, consecrating the great structure as an identifying element of Italian engineering. This route at some point follows a descending path

On June 30, 1964 the exhibition entitled "Art in a Changing World: 1884-1964: Twentieth Century Engineering" was inaugurated at the MOMA in New York, which enshrines the leading role of Italian engineering. Next to the already known works by Nervi, Morandi and Zorzi, a number of works by some of the protagonists of Italian construction were presented, unknown to the general public such as: Carlo Cestelli Guidi (1906-1995), Franco Levi (1914-2009), Arrigo Carré (1919 - 2004) and Giorgio Giannelli (1918 - 1985).

The gradual loss of centrality of the prestigious Istituto Sperimentale Modelli e Strutture (ISMES) where many tests and experimental models of constructive works were carried out, sometimes impossible to calculate with the usual rules of construction science and the Vajont disaster (1963), mark the point where the decline of this engineering parable begins, on which an inexplicable silence still descends today.

In 1997 at the Center Pompidou, a retrospective was set up named L'art de l'ingénieur, in which there still are to this day the admirable works of Nervi and Morandi to represent the Italian contribution during the twentieth century. The organization of the exhibition suggests that this heritage is not collected by any future generation , a victim of a failed process of historicization and therefore incapable of representing contemporaneity, represented instead by the tensile structures of Frei Otto or by Richard

IOP Conf. Series: Materials Science and Engineering **809** (2020) 012001 doi:10.1088/1757-899X/809/1/012001

Rogers high-tech and Norman Foster, and the constructive professionalism of the Ove Arup & Partners. [1]

Again in 2014, an exhibition inaugurated at MAXXI in Rome, offered a selection of studies, projects, models, drawings and photographs from a collection of 20th century engineers at the MAXXI Archives. On this occasion, among the various conferences, the " Scuola italiana di Ingegneria " by Sergio Poretti and Tullia Iori stood out, among other authors of the SIXXI research. Twentieth Century Structural Engineering: The Italian Contribution, in which they presented some results. [2].

2. The School of Italian Engineering

The protagonism of Italian engineering on the international scene has become lost over the years: the school of engineering has been somehow excluded from a historical perspective. Yet the extraordinary research on structural prefabrication, the testing of systems in pretensioned reinforcing tendons and the realization of slender vaults had led to a lasting success based on the expansive potential of the resulting structures.

To be able to resolve this issue, we should take a few steps back. The success of Italian engineering is undoubtedly based on the long experimentation and use of reinforced concrete, during the period of autarchy and until the Second World War. Reinforced concrete has almost completely replaced metal structures since the beginning of the 20th century. The pioneers in the scientific study and in the technical application of reinforced concrete structures are Camillo Guidi (1853 -1941) and Silvio Canevazzi (1852 - 1918), and then Arturo Danusso (1880 - 1968) and Gustavo Colonnetti (1886 - 1968)). The latter collaborated with the major Italian representatives of the Hennebique system. The collaboration between Danusso and Nervi, for example, has given rise to a whole series of applications on the slender vault, whose structural system gave the possibility of fulfilling the problem of low resistance to the breaking load of cement.

Danusso created the laboratory for models and constructions at polytechnic of Milan in 1931 and Ismes in Bergamo in 1951. From the collaboration with Nervi, the first celluloid model was released for the hangar project for the Air Force in Orvieto (date). In the same years, Nervi developed what would later be called Sistema Nervi, a new method for the production of reinforced concrete structures. The system offered a double advantage: on the one hand, to significantly reduce the use of construction formworks, on the other a respect for the material, based on prefabrication, assembly on site and ferrocement. This system, in its application variants, was perfect for large roofs. Still, the studies on prestressing gave the possibility to arrive at solutions that minimised the use of iron in the construction of reinforced concrete. Another researcher is associated with the use of this system Colonnetti, to whom we owe the projects of the first prestressed concrete bridges, between 1949 and 1951. These constructions were the final outcome of his research, first during the autarchic regime, then in exile in Lausanne, and finally as president of the CNR (National Research Center). [3 p. 120].

In the years that followed the Second World War, Italian engineering could apply much of the experimental research carried out up to that point on real structures. This constructive enthusiasm actually arose from a profound need for physical reconstruction and identity. These occasions gave the opportunity to introduce to the general public and not just a series of capable designers, who constituted a real generation: Nervi, Morandi, Krall, Cestelli Guidi, among the veterans and Zorzi, Musmeci, Carè and Giannelli, Galli and Franciosi, among the youngest. L'Autostrada del Sole, with its high number of bridges and viaducts, has given space for experimentation to the various personalities of the Italian School. In the same period Nervi, with his engineering and construction firm Nervi & Bartoli, imposed himself on the international scene with works designed and realized for the Olympic Games in Rome in 1960. On the other hand, in the same years, Riccardo Morandi developed an equally unique and characteristic design path.

This school has left a great deal of quality works, but apparently no heirs. Probably the first reflection on this decline should be looked for in the shipbuilding setting that has characterized the same Italian construction tradition, strongly artisan and traditional building. In fact, when the energy crisis of the seventies arrived, Italy's construction engineering witnessed a profound lack of preparation in the face of an increasingly conspicuous presence of high-tech engineering.

The tradition of the Italian shipyard is not associated with light materials and techniques, such as using glass and steel, a characteristic that has automatically excluded it from dialogue with contemporaneity. It will return to the fore only after a long time when, on the occasion of some great contemporary yards (the Roman church Dives in Misericordia by Richard Meier (1998-2003), the MAXXI by Zaha Hadid (1998-2009), the Italian Pavilion at the 2010 Expo in Shanghai) the need to demonstrate high competence in the cement and shipbuilding sectors applied to large structures returned. [1]

3. Intuition or science: the legacy of Nervi

Nevertheless, in this parable of oscillating dialogue with the contemporary and the definition of a trend or school, Pier Luigi Nervi continues to represent an incontrovertible model, even today. Its modernity, more than in the results, lies in the method, in which structure, form and space represent a unit. His contemporaneity lies in his constant, though sometimes unconscious, dialogue with the Italian cultural design past. Ernesto Nathan Rogers, in outlining a portrait of the engineer Nervi, compares it to Maillart, saying that as the latter he was able to give realism to the problem of tekne, bringing it back to its correct and original meaning. The work of these two engineers, Rogers continues, offers a new synthesis on the distinction between "science and art, between use and medium" which until then had always presented itself as a dichotomous binomial. It is the same synthesis that characterized creativity in the medieval architectural construction, in the hands of unconscious artisans, or that defined the architect in the Renaissance through the recovery of the classical approach, making it aware. Rogers goes on to say that in that synthesis the artistic vision of form and its material content must not be a twofold concept that generates divergent paths but rather an intense process in which both are interchangeable. When this process is the fruit of two opposing forces, the architect must solve them in a synthesis. Rogers stated that the word 'tekne' returns here to its original meaning and that it does not refer only, as in modern usage, to a coherent interaction of technique and execution at the service of science and art but, as the Greeks intended, to a synthesis of art, science, knowledge, craft, skill and profession. [4]

In an article in the magazine "L'Ingegnere" published perhaps fourteen years before Scienza o arte del costruire?, Nervi began to lay the foundations for reflecting on the empirical method of structural design, concepts repeatedly taken up in his later writings. The question that concerns the whole reflection is whether there can really be a true science of engineering, and how much truth exists and how much adherence to reality, in the numbers with which one dialogues with the resistance of materials and the forces acting on structures. He continues pondering what value these numbers that can be obtained after processes and formulas, which deal exactly with inaccurate things, can have. This need to transform approximation with accuracy leads to results that in one final phase require interpretation and how can we make this interpretation if not with a personal criterion, where the judgment parameter can only be intuition and feeling. In his view, the most serious damage of the scientific illusion of engineering lies in the consequent drying up of the imagination and of constructive intuition, which only, can create a beautiful, economic and stable work. And finally, he concludes in the text by saying that in the evaluation of non-measurable factors lies the true essence of engineering, but in this light it is no longer science but art. And when it refers to the latter term, it does so according to the meaning that Aristotle gave of it, namely mechane, which means ability combined with the capacity to grasp the physicality of reality, a characteristic of the practical intelligence (metis) of the engineer. If this does not get used to thinking and feeling the structure, and not simply to calculate it. [5]

You will continue to think that there will always be an incurable contrast between the mathematical mentality and the intuitive mentality, rich in aesthetic sensibility and creative imagination. [6]
Nervi has represented for Italian culture, building a milestone. His teachings have influenced, directly or indirectly, a vast culture of building. Around a kind of empiricism applied to engineering, a real school was established, although there never actually existed a conscious sense of belonging to a trend. Giorgio Morandi or later Sergio Musmeci, have participated in building a design attitude that very often makes us really reflect, looking at their constructions, if we are faced with works of "genius" or works of "art". It is in this dichotomous vision of construction that we should read the traces of Nervi's inheritance on his successors, who in their specificity have continued the work. (Figure. 1)

Musmeci, for example, as soon as he graduated he worked at the firm of Nervi & Bartoli, between 1949 and 1951, and then, in 1952, together with Antonio Nervi, started the studio di Architettura e Tecnica edilizia, with Nervi's father as a consultant. [7 p. 290]

Musmeci's attitude towards this diatribe between ingenuity and a work of art is clearly expressed on the occasion of the drafting of the text that he will write on the occasion of the death of the master. The great legacy that according to Musmeci, Nervi left to the culture of building were not so much its brilliant solutions but the problems left unresolved, because it is from here that one can start again and reflect on the "relations between the act of building and architectural culture on the one hand and the science of engineering on the other". [8 p. 12] Musmeci believed that, when it comes to sizing the sections of resistant structures, we have very sophisticated tools, when it comes to making the basic design choices, on which almost all of the result depends. It is in the same conditions as a Renaissance architect: "intuition, experience, static sense, but nothing that puts me in a position to choose (...) with awareness and responsibility". [9 p. 40]

This condition which for Nervi is an incontrovertible fact for Musmeci is unacceptable. For Nervi, this gap can contribute to shaping designers who, alongside the cold, impersonal formulation of the construction sciences, can combine intuition and imagination, giving life to a real "art of building". [10 p. 30]

If for Nervi the rigor coming from the highest mathematical procedures inhibits architectural invention, for Musmeci it is a question of inversion of terms. (Figure. 1)

That is, he maintains that the engineer must not start from the given form and limit himself to determining his tensions, but on the contrary he should be able to fix the acting forces and one or more performance parameters and consider the final shape of the structure as unknown. His design path, in fact, will always be set on finding the minimal form, in which each part is equally necessary. Obviously there is no single minimal form, but a set of possible forms that we could call structural organic forms; and it is within this circumscribed although rich group of possibilities that the intuition and creativity of the engineer are measured. [11 p. 2]

When Riccardo Morandi (1902-1989) began his professional career, Pier Luigi Nervi is already a famous engineer. Their relationship focuses on parallel paths that only cross a few occasions. Both are engaged in university teaching, the first in Florence, the second in Rome; both have their studio in Rome. They meet on some occasions like when in 1962 they were awarded by Aitec (Italian Association of Concrete Economic Technician). In 1960 they competed for the awarding of the works for the Olympics in Rome, but Nervi will build the Palazzetto dello Sport, the Corso Francia viaduct, the Flaminio stadium, the Palazzo all'Eur in the Olympic area, (Figure. 2) while Morandi will build the overpass. of Corso Francia and Fiumicino airport (in collaboration). [12 p. 262]

The works of the Olympics will make him a personality known internationally, while Morandi will remain on the fringes of fame, even when he built the 9-kilometer bridge on the Maracaibo lagoon in Venezuela in 1957. When Nervi passes away, Morandi entrusts his memory to the pages of the magazine "Industria Italiana del Cemento". It is in this essay that one can understand the deep bond that bound the two professionals. Referring to Nervi, Morandi describes him as one who has always influenced his professional life with the example of such coherent activity. Morandi states that he feels the need to analyse the reasons for his long spiritual bond with a man whom he has personally met only occasionally and with whom he has often been a competitor and sometimes also in polite controversy, this more or less clearly expressed. [13]

For Morandi, therefore, Nervi represents a model of professional ethics and coherence, an influence that will accompany him for a long time, despite their obvious distance in research and applications in the construction field. But despite a clear difference, there is a design attitude that unites them: leaving room for an unpredictability that can be solved through ingenuity and intuition.

When both are awarded the Aitec Prize by the Minister for Industry and Commerce, Emilio Colombo, in the hall of honor of the Palazzo della Civiltá e del Lavoro all'Eur, an exhibition is also set up of the awarded works. The judges reward the two masters for their unanimity, recognizing them as creators of an increasingly daring technique, to which concrete allows new processes, new dimensions and new architectural expressions, with a typical contribution to life, work and man's well-being. Colombo states Pier Luigi Nervi and Morandi have well deserved this recognition from their country because they have

honoured it in various world competitions, thus giving authority and prestige to the Italian stylistic and construction tradition. [12 p. 265]
When Morandi receives the telegram telling him that he was equally placed with the engineer Nervi, his satisfaction is doubled by the fact that he is united with a great personality like Professor Nervi.

4. Conclusion: a reflection as a new beginning?

The story of these correspondences, although limited by information available, opens up a series of questions not only linked to the success and decline of a School, but also to the broader concept of the transmission of knowledge and the recognition of a master, of a template. To this, we must add reflection whether or not there is a model of Italian knowledge. The characteristics of both the training and professionalism of Nervi, place him in a sequence of models recognized as emblematic for their typicality in the Italian tradition of building, but also of theoretical reflection around building. Rogers states that Nervi's secret lay in his ability to extract the essence of things and communicate it by expressing the truth. He does not create from nothing, but invents, giving accuracy and consistency to the term, used in its original meaning of invenio. Rogers states that it is like a medium that invokes the spirit of statics leading it to materialize; a spirit that others express all too often as incomprehensible mathematical symbols. The relationship between thought and creative activity is inextricably linked in his mind, continues Rogers, even if, in reality, there is nothing more difficult than expressing complicated things in a simple way. [4 p. 115]

Figure 1. Sergio Musmeci, Ponte sul Basento (Potenza), 1969.
https://focusarchitettura.wordpress.com/2014/07/27/strutture-romane-montuori-musmeci-nervi/

Nervi combines the two mental attitudes that distinguish those who create and construct spaces: the mathematical, analytical and synthetic, intuitive and artistic attitudes. These two models of being in the world complement each other, do not oppose or cancel each other out. [14]. And this makes it similar to a shared idea, even in an international context, that of Italianness. His example is not sectorial, but interdisciplinary and helps to overcome the ancient diatribe that poetry is entrusted to architects and prose to engineers; this fracture was partly overcome thanks to a way of making history that owes credit to Carlo Giulio Argan and Manfredo Tafuri.

But when a work of engineering like a bridge, recognized worldwide for its importance and for the ability to mark a surplus in the theory and building practice collapses, what does it drag with it? A personality, a profession, a model of knowledge, a cultural tradition?
A first reaction, from those who give a voice to culture, has been that of defense: defending the personality of the design engineer, removing responsibility, showing the absolute specificity of a modus of knowledge that has made us famous throughout the world, and to validate the validity of a tradition that comes from the distant Renaissance.

Now, with the silence of a memory that has already lasted a year, reflection brings to light a theme that increasingly belongs to the contemporary: that whatever object or thought is produced, the recipient always has the task of understanding its absolute preciousness and specificity, and to protect it and then be able to entrust it again.

IOP Conf. Series: Materials Science and Engineering **809** (2020) 012001 doi:10.1088/1757-899X/809/1/012001

Figure 2. Pierluigi Nervi: Palazzetto dello sport (1957) e Viadotto di Corso Francia a Roma (1960); Riccardo Morandi: Ponte sul Polcevera a Genova (1964).
https://focusarchitettura.wordpress.com/2014/07/27/strutture-romane-montuori-musmeci-nervi/

What we can conclude is that, as Argenti states, the Italian School of Engineering was not born as an ism and was eclipsed before the next formalist drift infected it. It was able to project itself forward, to be also a visionary, without ever having to sever its roots. And perhaps in the elegant sobriety of its structures, in the ability to connect technical knowledge to the creativity of design lies the most vivid and current part of its lesson. In this sense we can speak of a School. [15 p. 1]

5. References

[1] P. L. Nervi, «Corretto costruire,» Strutture, vol. 1, pp. 4-5, 1947.
[2] R. Capomolla, «Sergio Musmeci. L'ingegnere tra arte e scienza,» in Cantiere Nervi. La costruzione di una identitá. Storie, geografie, paralleli, Skira, A cura di, Milano, 2012, pp. 289-293.
[3] P. L. Nervi, «Scienza o arte dell'ingegnere?,» L'Ingegnere, vol. 7, n. 3, pp. 473-474, 1931.
[4] P. L. Nervi, Scienza o arte del costruire? Caratteristiche e possibilitá del cemento armato, Roma: Edizioni della Bussola, 1945.
[5] S. Musmeci, «Le tensioni non sono incognite,» Parametro, n. 80, pp. 36-40, 1979.
[6] S. Musmeci, «L'ereditá di Pier Luigi Nervi,» L'industria delle costruzioni, n. 3, pp. 5-20, 1979.
[7] S. Musmeci, «Ponte sul Basento a Potenza,» L'industria italiana del Cemento, n. 2, pp. 77-98, 1977.
[8] Santarelli, «Strutture romane. Montuori, Musmeci e Nervi,» 27 july 2014. [Online]. Available: https://focusarchitettura.wordpress.com/2014/07/27/strutture-romane-montuori-musmeci-nervi/ .
[9] E. N. Rogers, «A portrait of Pier Luigi Nervi,» in The works of Pier Luigi Nervi, E. N. Rogers, A cura di, New York, Frederick A. Praeger, 1957.
[10] T. Iori e S. Poretti, Twentieth Century Structural Engineering: SIXXI: The Italian Contribution, Roma: Gangemi, 2015.
[11] T. Iori e S. Poretti, «Storia dell'ingegneria strutturale italiana. Ascesa e declino,» Rassegna di architettura e urbanistica, n. 148, pp. 120-122, Gennaio-Aprile 2016.
[12] M. Argenti, «La Scuola italiana di ingegneria,» Rassegna di architettura e urbanistica, n. 148, Gennaio-Aprile 2016.
[13] M. Marandola, «Riccardo Morandi. I protagonisti antagonisti dell'ingegneria italiana del Novecento,» in Cantiere Nervi. La costruzione di una identitá. Storie, geografie, paralleli, Milano, Skira, 2012, pp. 262-266.
[14] R. Morandi, «Ricordo di Pier Luigi Nervi,» L'industria Italiana del Cemento, pp. 71-72, febbraio 1979.
[15] M. Salvadori, «Foreword,» in Structures, New York, F.W. Dodge Corporation, 1956, pp. V-VII.

A Fast Solving Method for Curve Profile of Suspension Bridge Catwalk System in Plane

Zhang Xiaoping[1,a], **Zeng Zhuo**[1,b] **and Li Zengzeng**[2,c]

[1] CCCC Wuhan Harbor Engineering Design and Research Co. Ltd., Wuhan, P. R. China
[2] China Communications 2nd Navigational Bureau 2nd Engineering Co. Ltd., Chongqing, P. R. China

[a]shoppinggre@hotmail.com, [b]390234103@qq.com, [c]1064167716@qq.com

Abstract. With increasing demands of in plane suspension bridge for mountainous area of P. R. China in recent year, rapid erection of catwalk system has become the researching hotspot for constructors, wherein curve profile is one of the key factors that matter. In this paper, a nonlinear optimization method is introduced for fast solving curve profile of catwalk system. To achieve the goal, statics fundaments and compliant characteristics of general catenary is firstly analyzed. Structural detailing of catwalk system is then described so that segmentation prior to computation can be made as per loads types or requirements of obtaining finer curve coordinates. In addition, nominal curve profile of catwalk system is derived according to segmentation of main cable, based on which the actual catwalk curve profile solving problem is afterwards formulated as an unconstrained nonlinear programming in terms of external penalty parameter and solved by using Levenberg-Marquardt (L-M) algorithm. Finally, a study case is shown to illustrate the effectiveness of proposed method.

1. Introduction

Catwalk, an indispensable structure for modern construction of suspension bridge, works as the temporary detours or footbridge for builders or other equipment to easily proceed wire wrapping, cable traction and installation into the saddle, etc. on both upstream and downstream sides of the bridge, and will be finally dismantled after completion of construction [1,2]. For accurate erection of main cable, catwalks shall be adaptively controlled as a similar parallel curve to main cables with certain clearance when main cables are in load-free condition before being fixed or anchored. As the transformation of catwalk curve profiles is executed by winch, pulley blocks and locking mechanism amid cable erection period [3], a fast guidance on dragging force, stress-free length, rope selection verification and its combination design, etc. are required [4].

Some researchers have been dedicated for decades on catenary analysis and its applications to in-plane suspension bridges based on cable elasticity theory[5-9], most of their works focus on solution of main cable curve profile and erection force prediction[10,11], wherein iteration algorithm like Newton-Raphson[2,7,12] and linearized computation techniques have been used[13-16]. With the development of computing tools, some other researches in terms of finite element method and indoor experiment [17-21] have been carried out to worked as evaluation measures not only for suspension bridge but all other cable supported ones. Whereas those methods are either accurate or ready-to-hand to some extent for design or verification of catwalk system, practicability for builders is relative less as they also have time-consuming demerit both for modeling and computation, and most importantly catwalk is a complex catenary structural system that identify itself to simply suspended cables. Compared to researches of main cable, majority works regarding catwalk system are focusing on formulating its erection technology and evaluation of wind resistance capability [22-26] rather than prediction curve profile, for which previous researches either regard rope combinations of catwalk platform and the same hanged on doorframes as a whole bearing structure but actually are separate controlled, or do not take handrails poles, wood transoms and other individually set facilities as point loads but transform them to uniformly distributed loads for computational convenience, or simply considering the objective catwalk curve profile as simply the translation of main cable curve profile that clearance in curve normal direction may vary Therefore, in this paper, we give general descriptions and design requirements of in-plane suspension bridge catwalk system, which is capable of considering all possible loader components as per required computation scale and propose a fast nonlinear optimization method for solving the curve profile by using L-M algorithm[27-29] for builder's better reference in practice.

2. Analysis of General Catenary

2.1Fundaments of Catenary.

As main cables and bearing rope combinations of catwalk platform or doorframes are either suspended between tower tops or ground anchoring structural components, the curve profile of which can be regarded as catenaries if only perfectly flexible assumption of cable is applied that action of any part of the line upon its neighbor is purely tangential, we can consider an arbitrary selected part of catenary at (x, z), show in Figure.1, and formulate its static equilibrium equations as:

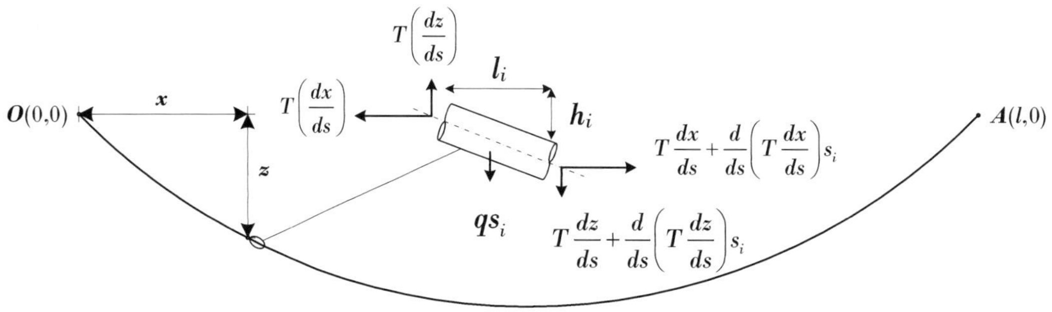

Figure.1 Statics Analysis of a General Catenary.

$$\frac{d}{ds}\left(T\frac{dz}{ds}\right) = -q \tag{1}$$

$$\frac{d}{ds}\left(T\frac{dx}{ds}\right) = 0 \tag{2}$$

where T is the tension of cable, s is the length of cable, q is the cable weight per unit length. dz/ds and dx/ds are the sine and cosine of inclination angle.

As equation (2) may be integrated to

$$T\frac{dx}{ds} = H \tag{3}$$

Equation (1) can be further rewritten as

$$H\frac{d^2z}{dx^2} = -q\frac{ds}{dx} \tag{4}$$

where H is the constant horizontal component of cable tension T, since there is no longitudinal force acting on rope along its curve.

Given that

$$\left(\frac{dx}{ds}\right)^2+\left(\frac{dz}{ds}\right)^2=1 \tag{5}$$

then differential equation of rope element may be finalized as

$$H\frac{d^2z}{dx^2}=-q\left\{1+\left(\frac{dz}{ds}\right)^2\right\}^{1/2} \tag{6}$$

the solution of which can be described in terms of Lagrange coordinate s as follow

$$x(s)=\frac{H}{q}\left\{sh^{-1}\left(\frac{V}{H}\right)-sh^{-1}\left(\frac{V-qs}{H}\right)\right\} \tag{7}$$

$$z(s)=\frac{H}{q}\left\{\sqrt{1+\left(\frac{V}{H}\right)^2}-\sqrt{1+\left(\frac{V-qs}{H}\right)^2}\right\} \tag{8}$$

where V are the vertical component of rope tension.

In addition, we may regard this arbitrary part as the ith segment of cable, $i=1,2,\ldots,m$, m is numbers of segments as per requirements of modeling and computation complexity required, we obtain distance components l_i and h_i between the two ends of the ith segment in horizontal and vertical directions,

$$l_i=\frac{H_i}{q}\left\{sh^{-1}\left(\frac{V_i}{H_i}\right)-sh^{-1}\left(\frac{V_i-qs_i}{H_i}\right)\right\} \tag{9}$$

$$h_i=\frac{H_i}{q}\left\{\sqrt{1+\left(\frac{V_i}{H_i}\right)^2}-\sqrt{1+\left(\frac{V_i-qs_i}{H_i}\right)^2}\right\} \tag{10}$$

and further have cable length s_i by rewriting equation (9)

$$s_i=\left\{V_i-sh\left[sh^{-1}\left(\frac{V_i}{H_i}\right)-\frac{ql_i}{H_i}\right]H_i\right\}/q \tag{11}$$

When $|T(s)/EA_0|\ll 1$, stress-free length of arbitrary segment of cable can be also described as

$$s_{i0}=s_i-\frac{H_i}{2EA_0c_a}\left[c_al_i+sh\left(2c_al_i+c_b\right)-sh\left(2c_b\right)\right] \tag{12}$$

where $c_a=q/H_i$, $c_b=sh^{-1}\left\{h_ic_a/2sh\left(c_al_i/2\right)\right\}-c_al_i/2$. E is the elastic modulus, A_0 is the equivalent section area of cable. Further, we consider a vertical point load P_i exerted at the end of the ith segment, the second consecutive segment $(i+1)$ may have

$$H_{i+1}=H_i=H \tag{13}$$

$$V_{i+1}=V_i-\left(P_i+qs_i\right) \tag{14}$$

2.2 Compliant characteristics of catenary.

In order to solve the curve profile of catenary segment, constant updated linearized compliant relationship between tension and curve profile is better required for solution of nonlinear optimization programming. From equation (9) and equation (10), we have,

$$d\mathbf{c}_i=\begin{bmatrix}dl_i\\dh_i\end{bmatrix}=\begin{bmatrix}J_{l,H}&J_{l,V}\\J_{h,H}&J_{h,V}\end{bmatrix}\begin{bmatrix}dH_i\\dV_i\end{bmatrix}=\mathbf{J}_c d\mathbf{T}_i \tag{15}$$

Where $\mathbf{J}_c$ is the Jacobian matrix that represent the linear compliance characteristics of each cable segment, components of which can be described as

$$J_{l,H} = \frac{1}{q}\left\{ sinh^{-1}\left(\frac{V_i}{H_i}\right) - sinh^{-1}\left(\frac{V_i - H_i}{H_i}\right)\right\} + \frac{1}{q}\left\{ -\frac{V_i / H_i}{\left[1+(V_i / H_i)^2\right]^{1/2}} + \frac{(V_i - qs_i)/H_i}{\left[1+((V_i - qs_i)/H_i)^2\right]^{1/2}}\right\} \tag{16}$$

$$J_{l,V} = \frac{1}{q}\left\{\left[1+\left(\frac{V_i}{H_i}\right)^2\right]^{-1/2} - \left[1+\left(\frac{V_i - qs_i}{H_i}\right)^2\right]^{-1/2}\right\} \tag{17}$$

$$J_{h,H} = \frac{1}{q}\left\{\left[1+\left(\frac{V_i}{H_i}\right)^2\right]^{1/2} - \left[1+\left(\frac{V_i - qs_i}{H_i}\right)^2\right]^{1/2}\right\} + \frac{2}{q}\left\{ -\frac{V_i^2 / H_i^2}{\left[1+(V_i / H_i)^2\right]^{1/2}} + \frac{\left[(V_i - qs_i)/H_i\right]^2}{\left[1+((V_i - qs_i)/H_i)^2\right]^{1/2}}\right\} \tag{18}$$

$$J_{h,V} = \frac{1}{q}\left\{\frac{V_i / H_i}{\left[1+(V_i / H_i)^2\right]^{1/2}} - \frac{(V_i - qs_i)/H_i}{\left[1+((V_i - qs_i)/H_i)^2\right]^{1/2}}\right\} \tag{19}$$

3. Detailing of Catwalk System and Computation Preparation of Curve Profile

3.1. Structural Composition.

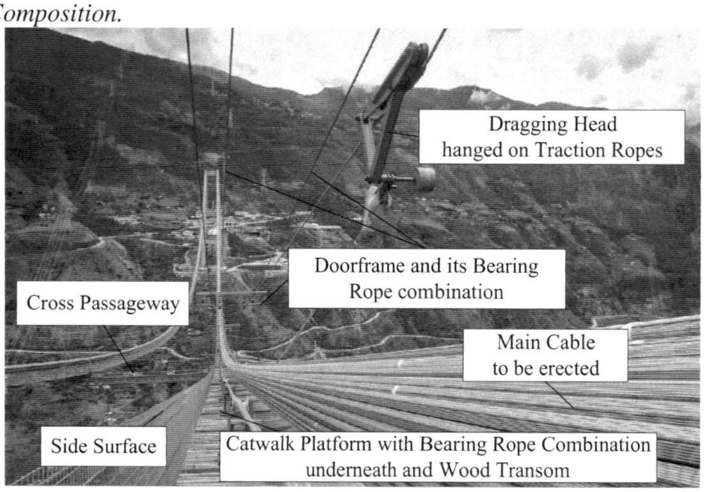

Figure. 2 General Catwalk Structural Components.

As a very important bearing structure and transportation platform for main cables erection, see Figure.2, catwalks system generally consists of manipulation platform, doorframes, cross passageways, and is often supported in terms of bearing ropes combinations suspended on tower top or fixed to ground anchoring structures on both ends of bridge span. There are two main bearing rope combinations for erection of catwalk system, one is usually set underneath main cable curve profile with a constant 1.2*m*-1.5*m* clearance and is together with uniformly arranged bottom and side net surface plates, wood transoms, handrails and its supporting poles to form a workable bench for builders manipulation convenience and passage of wrapping machine, the other is set mainly for force distribution of catwalk system and proper erection of discretely arranged doorframes and cross passageways. Generally, catwalk platform width is accordingly designed as per requirements of work space where 4-5*m* width is enough for most of the cases unless otherwise required to increase to obtain more stability capacity to counteract the heavy wind effects, in condition of which measures using extra wind cable combinations or increasing numbers of cross passageways shall be taken when wind load prevails.

3.2 Segmentation of Catwalk System.

Unlike main cable in load-free condition, which can be simply regarded as a catenary, two rope combinations of catwalk system bear impacts of uniformly distributed loads from various kind of functioning ropes assembled on and regularly arranged point loads from structural components or equipment in a certain form of force distribution at locations of doorframes. Thus, to decouple this complex relationship and to have a clear guidance for iterated solving curve profile, we reciprocally divide main cable and catwalk system along longitudinal direction into m segments $\mathbf{l}=[l_1,...,l_m]^T_{m\times1}$ as per above description of structural compositions, requirements computational complexity or obtaining more refined curve coordinates.

4. Nominal Curve Profile of Catwalk System

4.1 Formation of Main Cable Curve Profile Solving Problem.

As to acquire the full curve profile of catwalk system is the target of this paper, main cable curve profile shall be obtained in prior by using aforementioned catenary analysis as per identical segmentation in bridge longitudinal direction so that each nominal coordinates of catwalk platform can be available by then as computation reference of actual curve profile.

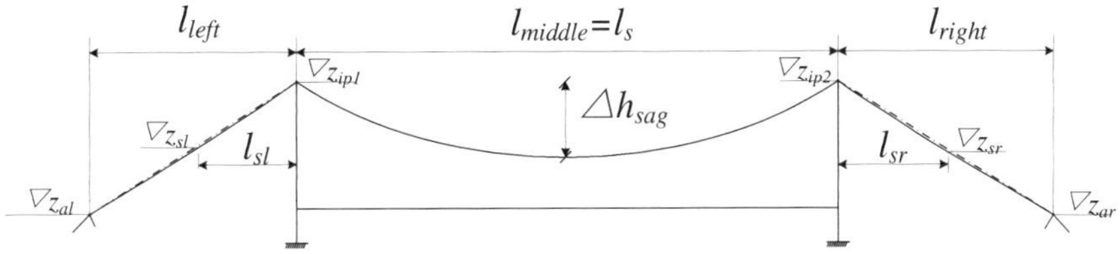

Figure.3 Design of Main Cable in Load-Free Condition.

As shown in Figure.3, we general consider the middle span for instance, wherein the span length l_s, level difference $\triangle h = z_{ip1}$-z_{ip2} between main cable intersection points (IP) on tower saddles, sag values $\triangle h_{sag}$ at middle point span are given as the design inputs, and assume the segmentation of main cable as $\mathbf{l}_{cable}=[l_1,...,l_{n-1},l_n,...,l_m]^T$, $sum(\mathbf{l}_{cable})=l_s$, where l_{n-1} and l_n are the horizontal lengths of left and right segments beside middle point respectively. Generally speaking, n = $m/2$. Then, we can sequentially compute every curve profile in terms of recursive using equations (9), (10) and equation (13), (14) if tension vector $\Omega_{cable}=[H_{ip1},V_{ip1}]^T$ at the staring IP is regarded as the design variables to be solved. Tension at the ending IP $T_{ip2}=[H_{ip2},V_{ip2}]^T$ may be carried out as well by using equation (13) and (14) one more time. Thus, we can formulate main cable curve profile solving problem as following nonlinear programming problem:

$$minimize\ \Psi(\Omega) \tag{20}$$

where $\Psi(\Omega)=[\varphi_1^T,\varphi_2^T]^T$ is the objective function vector, $\varphi_1=\sum_{i=1}^{n-1}h_i-\Delta h_{sag}$, $\varphi_2=\sum_{i=1}^{m}h_i-\Delta h$.

4.2 Solution of nominal catwalk system curve profile.

In case of the main cable curve profile is known, as shown in Figure.4, we may consider the ith cable segment start from point $\mathbf{p}(x_i,z_i)$ for an assumed horizontal length l_i, the vertical clearance to point $\mathbf{p}_{ci}$ on catwalk system shall be

$$h_i^{'}=h_i+C\cos\beta_{i+1} \tag{21}$$

where h_i, β_{i+1} are corresponding vertical lengths of the ith segment and inclination angle of the $(i+1)$th segment which can be solved in terms of given clearance C, previously defined equations (9), (10), (13), (14) and following equations (22) and (23).

$$\beta_{i+1}=\tan^{-1}\left(\frac{V_{i+1}}{H_{i+1}}\right) \tag{22}$$

$$l_i=C\sin\beta_{i+1} \tag{23}$$

Then, we can recursively obtain all nominal curve profile parameters $\mathbf{l}_{\text{ncat}}$ and $\mathbf{h}_{\text{ncat}}$ of catwalk platform, as well as the rope combinations of doorframes $\mathbf{l}_{\text{ndf}}$ and $\mathbf{h}_{\text{ndf}}$, at corresponding segment point $\mathbf{p}_{ci}\left(x_i, z_i - h_i^{'}\right)$ which are all described in the identical Cartesian coordination system.

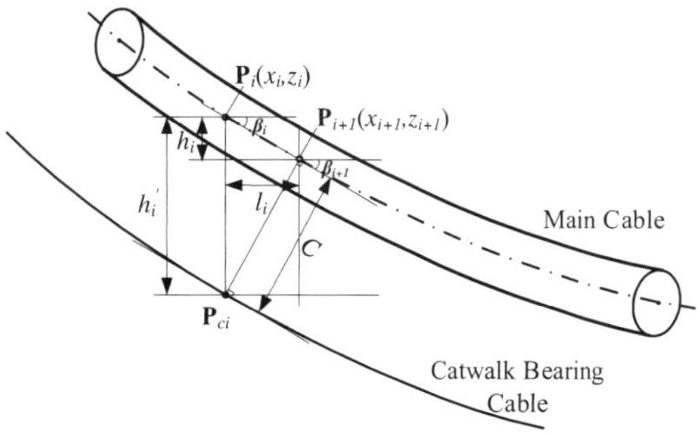

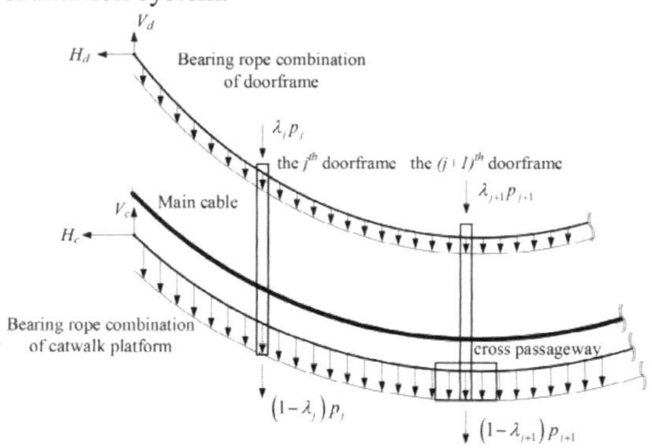

Figure.4 Solution of Nominal Catwalk Curve Profile. **Figure.5** Sketch of the jth Force Distribution

4.3 Force distribution on doorframes.

As rope combinations underneath catwalk platform and the same hanged on doorframes simultaneously bear whole loads of catwalk system in an uncertain distributed manner through doorframe structure, please see Figure.5, we assume there are u numbers of doorframes wherein the jth interaction between two bearing rope combinations located at the jth doorframe to be

$$p_j = p_{jc} + p_{jd} = \left(1-\lambda_j\right) p_j + \lambda_j p_j \tag{24}$$

where p_j is sum of the point loads exerted by doorframe or cross passageway at location of the jth doorframe, $p_{jc} = \left(1-\lambda_j\right) p_j$ and $p_{jd} = \lambda_j p_j$ are the force components respectively supported by bearing rope combinations underneath manipulation platform and hanged on the jth doorframe, $\lambda_j \in [0,1]$ is the jth force distribution factor to be solved.

5. Optimal Solution of Catwalk System Curve Profile

5.1 Formulation of Catwalk System Curve Profile Solving Problem.

Using catenary statics, curve profile solutions of main cable and nominal catwalk system, force distributions relationship at each doorframe, we can formulate the catwalk system curve profile solving problem as the following constrained nonlinear programming problem

$$\textit{minimize}\ \Xi(\Omega),\ \textit{subject to}\ D: \begin{cases} \mathbf{g}_j(\Omega) = 0,\ j = 1,\dots,u+1 \\ \mathbf{h}_k(\Omega) \le 0,\ k = 1,\dots,2u \end{cases} \tag{25}$$

where $\Omega = \left[H_{c1}, V_{c1}, H_{d1}, V_{d1}, \lambda_1, \dots, \lambda_u\right]^T_{1\times(4+u)}$ is the design variables, $H_{c1}, V_{c1}, H_{d1}, V_{d1}$ are the tension components of bearing rope combinations for catwalk system and doorframe at starting point. $\Xi(\Omega) = \left[\delta_1(\Omega), \dots, \delta_m(\Omega)\right]^T_{m\times1}$, $\mathbf{g}_j(\Omega) = \left[g_1(\Omega), \dots, g_{u+1}(\Omega)\right]^T_{(u+1)\times1}$ and $\mathbf{h}_k(\Omega) = \left[-\lambda_1, \dots, -\lambda_u, \lambda_1 - 1, \dots, \lambda_u - 1\right]^T_{2u\times1}$ are the objective function and boundary conditions. $\delta_i(\Omega) = \sum_{i=1}^{m}\left(h_{ic} - h_{inc}\right)$ and $g_j(\Omega) = \sum_{j=1}^{u}\left(h_{jd} - h_{jnd}\right)$ represent the ith and jth accumulative deviations of nominal curve profile to actual values of catwalk platform and doorframe rope combinations at segment points, respectively. $h_k(\Omega)$ represents all inequality constraints of force distribution control at location of each doorframe.

Accordingly, we can transform this constrained nonlinear programming into an unconstrained nonlinear programming problem in terms of introducing a very large penalty parameter r that the minimum of the problem (26) corresponds to the solution of problem (25).

$$minimize\ \Pi(\Omega) = \Psi^T(\Omega)\Psi(\Omega) \tag{26}$$

where $\Psi(\Omega) = \left[\Xi^T(\Omega), r^{1/2}\mathbf{g}_j^T(\Omega), r^{1/2}\mathbf{h}_k^T(\Omega)\right]^T = \left[\psi_1^T(\Omega), \dots, \psi_{m+3u+1}^T(\Omega)\right]^T$, the gradient vector of which can be given as $\nabla\Pi(\Omega) = 2\mathbf{J}^T(\Omega)\Psi(\Omega)$. Then, Hessian matrix $\nabla\Pi^2(\Omega) = 2\mathbf{J}^T(\Omega)\mathbf{J}(\Omega) + 2\mathbf{G}(\Omega)$ can be obtained, wherein $\mathbf{G}(\Omega) = \sum_{i=1}^{m+3u+1} \psi_i(\Omega)\nabla^2\psi_i(\Omega)$.

5.2 Application of Levenberg-Marquardt Algorithm.

Levenberg-Marquardt algorithm is an efficient and effective method that providing numerical solutions for nonlinear minimization. By modifying parameters during iterations, this algorithm can improve the disadvantages of Gauss-Newton algorithm and settle problems when initial value is too far from the local minimum solution. Thus, in terms of using L-M algorithm, we can firstly determine the search direction $\theta_{(t)}$ for the tth iterative numerical computation through following equation (27)

$$\left(\mathbf{J}_{(t)}^T\mathbf{J}_{(t)} + \upsilon_{(t)}\mathbf{I}\right)\theta_{(t)} = -\mathbf{J}_{(t)}^T\Psi_{(t)} \tag{27}$$

where $\mathbf{J}_{(t)}$, $\Psi_{(t)}$ are the Jacobian matrix and function vector at the tth iteration. $\upsilon_{(t)} \geq 0$ and $\mathbf{I}$ are scalar member and identical matrix, respectively.

Function $\Psi_{(t+1)}$ and its design variables $\Omega_{(t+1)}$ can be further formulated as

$$\Pi_{(t+1)} \approx \Pi_{(t)} + \alpha_{(t)}\theta_{(t)}^T\nabla\Pi\left(\Omega_{(t)}\right) \tag{28}$$

$$\Omega_{(t+1)} = \Omega_{(t)} + \alpha_{(t)}\theta_{(t)} \tag{29}$$

where $\alpha_{(t)} \geq 0$ is the step length of the tth iteration. Further, we rewrite equation (27) into equation (30) as

$$\theta_{(t)}^T\nabla\Pi\left(\Omega_{(t)}\right) = -2\theta_{(t)}^T\left(\mathbf{J}_{(t)}^T\mathbf{J}_{(t)} + \upsilon_{(t)}\mathbf{I}\right)\theta_{(t)} \tag{30}$$

when $\upsilon_{(t)}$ is sufficiently enough, $\theta_{(t)}^T\left(\mathbf{J}_{(t)}^T\mathbf{J}_{(t)} + \upsilon_{(t)}\mathbf{I}\right)\theta_{(t)}$ is positive definite that $\theta_{(t)}^T\nabla\Pi\left(\Omega_{(t)}\right) < 0$, which draws the conclusion that

$$\Pi_{(t+1)} < \Pi_{(t)} \tag{31}$$

So that, we can finally obtain the optimal solution of cable curve profile as per specified convergent indexes.

6. Case Study

In this section, an engineering practice of catwalk under construction is shown to verify the effectiveness of this method.

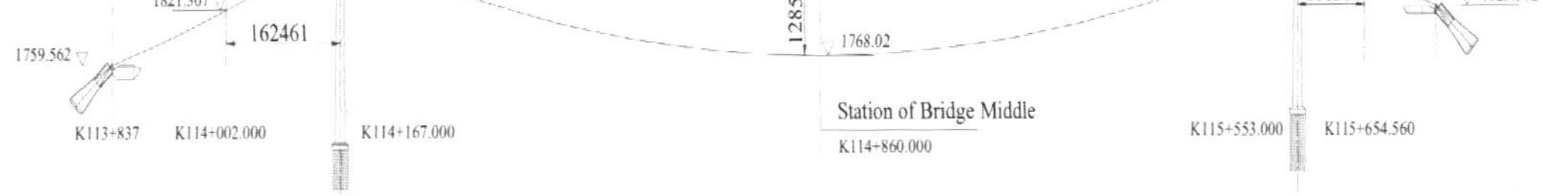

Figure.6 Design of Main Cable in Load-Free Condition for an Under-Construction Suspension Bridge.

As shown in Figure.6, the span layout of suspension bridge is 330*m*+1386*m*+205*m*, curve profile of main cable, with 21.7*kg*/*m* per rope in weight, in load-free condition, wherein sags for each span is designed as 6.755*m*, 128.542*m* and 2.599*m*. The corresponding catwalk platform, controlled to be 1.5*m*/1.7*m* below main cable curve profile in middle span and side spans, uses 10-Ø54*mm* and 2-Ø54*mm* steel wired ropes, each one of which is $12.2kg/m$ in weight and with $1960MPa$ tensile strength grade, to form bearing rope combinations respectively for itself and proper erection of doorframes of $7m$ in height. Doorframes and cross passageways are set along bridge direction every $57.75m$ and $173.25m$ for middle span, and about every $50m/55m$ and $100m/150m$ for side spans, based on which segmentations vectors along longitudinal direction are carried out as:

$l_l = \left[52.461,\ 55,\ 55,\ 50,\ 50,\ 67.539\right]^T_{6\times1}$, $l_m = \left[57.75,...,57.75\right]^T_{24\times1}$ and $l_r = \left[51.564,\ 50,\ 50,\ 53.436\right]^T_{4\times1}$

where l_l, l_m and l_r are in meters, the design of which may be finer as per user required computation scale. In addition, loads applying of catwalk system are listed in Table 1, wherein, uniformly distributed loads (*U.D.L.*) and point loads (*P.L.*) are comprehensively considered.

Table 1. Loads Applying of Catwalk System

Category	Items	Load/Unit	Quan.
U D L.	Catwalk platform rope	12.2 *kg/m*	10
	Handrail rope	3.81 *kg/m*	2
	Doorframe rope	12.2 *kg/m*	2
	Traction rope	5.42 *kg/m*	2
	Surfaces	58.51 *kg/m*	—
	Ancillary	10 *kg/m*	—
	Pedestrian	5 *kg/m*	—
P. L.	Doorframe	250 *kg*	23
	Passageway	6348.2 *kg*	7

By adopting computation procedure, we first obtain the curve profile of main cable, the same of nominal catwalk system and doorframe rope combination, which is shown in Figure.7 with data aspect ratio of 1:1, as the computation reference. Then, we can correspondingly figure out the curve profile deviations, point load distribution factors, tension forces, free of stress rope length of both bearing rope combinations in each span in terms of using our proposed method, which are respectively shown in Figure.8, Figure.9, Figure.10 and Table 2.

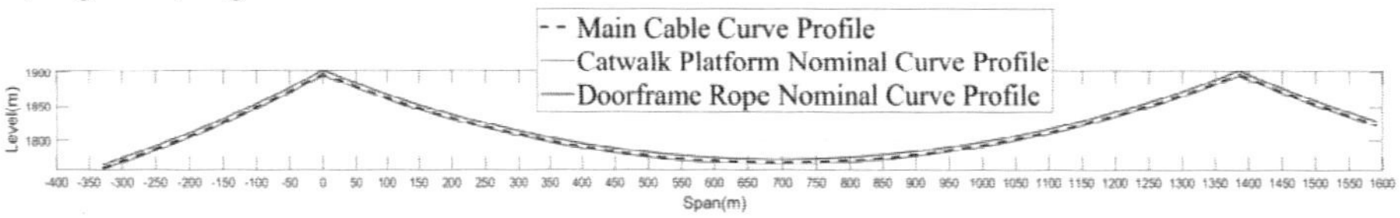

Figure.7 Curve Profile of Main Cable, Catwalk Platform and Doorframe Rope Combinations.

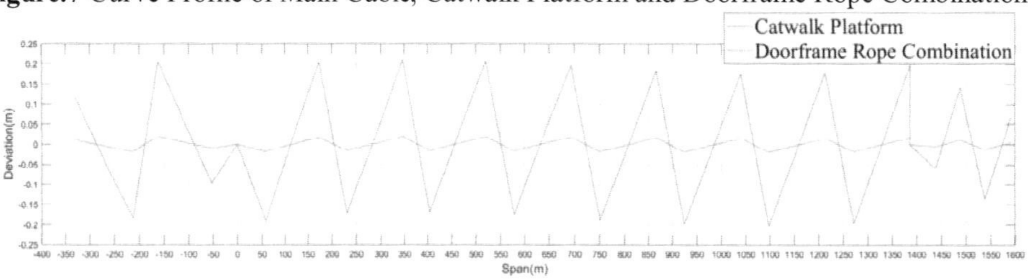

Figure.8 Curve Profile Deviations of Catwalk Platform and Doorframe Rope Combinations.

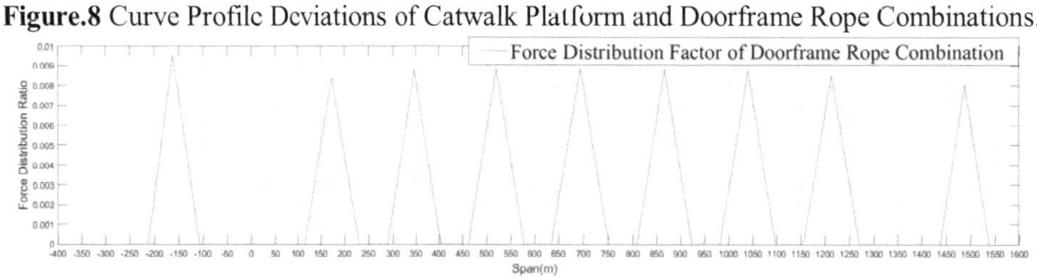

Figure.9 Force Distribution at Location of Each Doorframe.

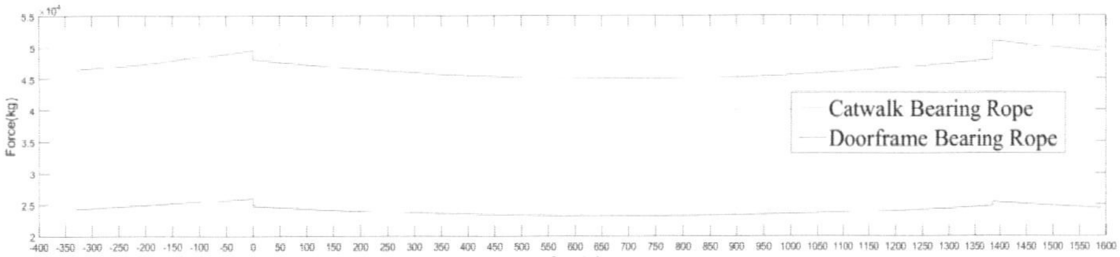

Figure.10 Tensions of Single Bearing Rope for Catwalk and Doorframe.

Table 2 Free of Stress Length of Ropes

Items\Span	Left	Middle	Right	Sum
Main Cable	357.41 *m*	1416.24 *m*	217.53 *m*	1991.18 *m*
Catwalk rope	356.72 *m*	1413.61 *m*	217.08 *m*	1987.41 *m*
Doorframe rope	357.75 *m*	1417.10 *m*	217.72 *m*	1991.57 *m*

The computation results show that the curve profile deviation of catwalk platform is controlled within 0.2*m*, which is a proper result for builder's manipulation in practice, while the same of doorframe rope is relatively less than 0.05*m* that most of the loads are supported by catwalk platform bearing rope, maximum tension of which is 5.1e4*kg* and occurred at right IP with safety factor of 4.06. Moreover, free of stress rope lengths are also available through this computation for builder's proper preparation and implementation.

7. Conclusion

To conclude, this paper mainly focused on providing a fast solving method for prediction of catwalk system curve profile amid main cable erection in terms of using L-M algorithm, which may work as a replacement of the complex and time-consuming finite element methodology for builders and designers. To achieve that, we firstly analyzed the compliant characteristics of general suspended cable. In addition, by description of catwalk system, we correspondingly divided the main cable and catwalk system for convenience of computation. Then, we formulated main cable curve profile solving problem and gave the solution of nominal curve profile of catwalk system, based on which the numerical solution of actual curve profile can be derived by using the L-M algorithm. Moreover, a study case of catwalk design for a under construction suspension bridge was shown for further illustration of proposed method, the results of which have showed the effectiveness of this proposed method.

8. References

[1] H. M Irvine, in: Statics of suspended cables, J. Engrg. Mech. Div., ASCE, vol. 101-EM13 (1975), pp. 187-205

[2] A. H. Peyrot and A. M. Goulois, in: Analysis of cable structures, Computers and Structures, vol. 93-ST1 (1967), pp. 1-26

[3] SA. Saafan, in: Theoretical analysis of suspension bridge, J. Struct. Eng. Div., ASCE, vol. 109 (1966), pp. 1952-1971

[4] Y. M. Desai, A. H. Shah and D. N. Buragohain, in: Geometric nonlinear static analysis of cable supported structures, Computers & Structures, vol. 29-6 (1988), pp. 1001-1009

[5] D. M. Brotton, in: The solution of suspension bridge problems by digital computers—Part I, The structure Engineer, vol. 41-4 (1963), pp. 121-126

[6] K. Ahmadi-Kashani and A. J. Bell, in: The analysis of cables subject to uniformly distributed loads, Engineering Structures, vol.10 (1988), pp. 174-184

[7] D. M. Brotton, in: A general computer program for solution of suspension bridge problems, The structure Engineer, vol. 44-59 (1966), pp. 161-167
[8] T. W. O'Brien, in: General solution of suspended cable problems, J. Struct. Eng. Div., ASCE, vol.93-ST1 (1967), pp. 1-26
[9] T. W. O'Brien and A. J. Francis, in: Cable movements under two-dimensional loadings, J. Struct. Eng. Div., ASCE, vol. 90-ST3 (1964), pp. 89-123
[10] C. A. Diego and C. A. Angel, in: Preliminary static analysis of suspension bridge, Engineering Structure, vol. 23 (2001), pp. 1096-1103.
[11] L. C. Fan, Y. R. Pan and G. H. Du, in: Study on the fine method of calculating the erection-parameters of long-span suspension bridges, CHN. J. Civil Eng., Vol. 32-6 (1999), pp. 20-25
[12] H. M. Irvine, in: *Cable Structure*, MIT press, 1981.
[13] Y. F. Zhou and S. Chen, in: Iterative nonlinear cable shape and force finding technique of suspension bridges using elastic catenary configuration, J. Eng. Mech., ASCE., vol. 145-5 (2019), pp. 04019031-04019031
[14] H. B. Jayaraman and W. C. Knudson, in: A curved element for the analysis of cable structure, Computers & Structures, vol.14 (1981), pp. 325-333
[15] G. P. Wollmann, in: Preliminary analysis of suspension bridges, J. Bridge Engrg., ASCE, vol.6 (2001), pp. 227-233
[16] W. M. Zhang, T. Li and L. Y. Shi, etc., in: An iterative calculation method for hanger tensions and the cable shape of a suspension bridge based on the catenary theory and finite element method," J. Adv. Struct. Eng., vol. 22 (2019), pp. 1566-1578
[17] R. Karoumi, in: Some modeling aspects in the nonlinear finite element analysis of cable supported bridges, Computers & Structures, vol.71 (1999), pp. 397-412
[18] H. Ozdermir, in: A finite element approach for cable problems, Int. J. Solids Struct., vol.15 (1979), pp. 427-437.
[19] F. Isaac, in: Large deformation static and dynamic finite element analysis of extensible cables, Computers & Structures, vol. 15 (1982), pp. 315-319
[20] N. J. Gimsing, in: *Cable supported bridges-concept & design*, 2nd ed. Chichester: John Wiley & Sons Ltd., (1997), in press.
[21]J. C. Russell and T. J. Lardner, in: Statics experiments on an elastic catenary, J. Engrg. Mech., ASCE, vol. 123 (1997), pp. 1322-1324
[22] P. Guo, C. Q. Wang and Y. A. Hu, etc., in: Influence of catwalk design parameters on the galloping of constructing main cables in long-span suspension bridges, J. Vibro. Eng., Vol. 19 (2019), pp.4671-4684
[23] X. P. You, Z. X. Zhang and Q. T. Yao, in: Line-shape calculation method for suspension bridge catwalk considering different states of bearing cables, CHN. J. Eng. Sci., vol. 10(2012), pp.12-15
[24] M. Liu, G. B. Xia and L. J. Liu, etc., in: The key measures on safety and linear control in catwalk erection, J. Applied Mech. & Materials, vol. 3843(2015), pp.837-844
[25] J. Wang, Y. Li and J. X. Liu, etc., in: Nonlinear aerostatic response of catwalk of suspension bridge, CHN. J. Highway and Trans. R & D (Eng. Ed.), vol. 7-2 (2013), pp.40-45
[26] C. A. Diego and C. A. Angel, in: Improving the wind stability of suspension bridge during construction, J. Struct. Eng. Div., ASCE, vol. 127-8 (2001), pp. 869-875
[27] R. L. Fox, in: *Optimization Methods for Engineering Design*. Reading, MA: Addison-Wesley, (1971), in press
[28] L. E. Scales, in: *Introduction to Non-Linear Optimization*. New York: Macmillan, (1985), in press.
[29] K. Levenberg, in: A method for the solution of certain non-linear problems in least squares, Quarterly of Applied Mathematics, Vol. 2-2 (1944), pp. 164-168

IOP Conf. Series: Materials Science and Engineering **809** (2020) 012003 doi:10.1088/1757-899X/809/1/012003

The School-building Model of the Church Schools in Nanjing, China before the 20th Century——Taking the Huiwen Academy as an Example

Hechi Wang

Hubei University of Technology, Wuhan, China.

wanghechi@163.com

Abstract. Church school is a special type of school in old China. After the first Opium War, Western churches have come to China to preach. Missionaries open churches, schools, and hospitals in China. This paper selects the largest church school in Nanjing in the late Qing Dynasty(before 1911) as an example. Based on the historical data of archival materials and the research methods of on-site mapping, the initial campus construction model of Nanjing Church School is summarized. The article first introduces the background of the school in China, the reasons for the establishment of the school, the history of the school; then analyzes the location, construction process, campus planning and architectural characteristics of the school; further analyzes the reasons why the western church replicates the western campus construction model; Finally, through the development of

IOP Conf. Series: Materials Science and Engineering **809** (2020) 012003 doi:10.1088/1757-899X/809/1/012003

educational buildings, the contributions of church schools in the history of modern education in China are summarized. The article shows the development of large-scale educational buildings that are rare in the history of modern Chinese education. It not only clearly clarifies the history of the development of church schools in China, but also has important guiding significance for the heritage protection of today's historical buildings. At the same time, it also has important reference value for the protection of educational building heritage and campus construction in China and even in the world.

1. School history

The site of Huiwen Academy (now Jinling Middle School) is located at No. 169 Zhongshan Road, Gulou District, Nanjing. It was founded in 1888 by the US Christian missionary Fowler (C.H.Fowler) on the Ganhe River in Nanjing. The English name is "The Nanking University" and the American missionary Ferguson (J.C.Ferguson) is the first dean[1] (Figure 1). It was first higher education institution established by Western missionaries in Nanjing[2]. This group of buildings was built in one time at the end of the 19th century. It is the largest church school in Nanjing before the 20th century. The school is grand in scale, elegant in architecture, and has high artistic value. It has witnessed the development of Nanjing Mission School .Its Bell tower is the first three-storey building in Nanjing in the past 100 years(Figure 2). It is still in use today and is the symbol of Nanjing education.

Figure 1. J.C. Ferguson.
source: Yale University Library, USA

Figure 2. Students in the classroom of the Bell tower (1909)
source: Yale University Library, USA

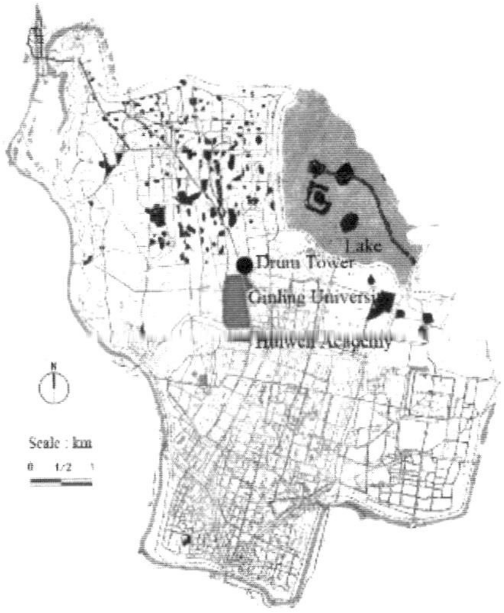

Figure 3. School site
source: Wisconsin Historical Society, USA

2. School construction

2.1 Site selection and construction process

The Huiwen Academy was located in the middle of the Nanjing (Figure 3), and its location was related to the scope of the missionary's activities at the time. The original missionaries preached in the four poles (now Mochou Road) and the Drum Tower along the riverside. They built churches, opened hospitals, and set up schools. The centralized setting of church buildings was conducive to the formation of a religious cultural circle.

The campus planning and architectural design of the Huiwen Academy was

IOP Conf. Series: Materials Science and Engineering **809** (2020) 012003 doi:10.1088/1757-899X/809/1/012003

completed by an American architect and Chen Mingji was constructed[1].

2.2 Campus planning

At the end of the19th century, the American church hired American architects to complete the campus planning and design of the Huiwen Academy (Figure 4),In the decades after the school construction, the planning layout of the Huiwen Academy was not changed. In 1910, Hongyu Academy was merged into the Huiwen Academy, and the merged group was built into Ginling University. The school purchased 156 Hectare of land as the new campus of Ginling University. The Huiwen Academy was the affiliated middle school of Ginling University and renamed Jinling Middle School. Therefore, in the overall campus planning of Ginling University, only the Huiwen Academy as the middle school of Ginling University, the environment and roads were partially adjusted and improved. The original campus planning of the Huiwen Academy was not adjusted. (Figure5,6)

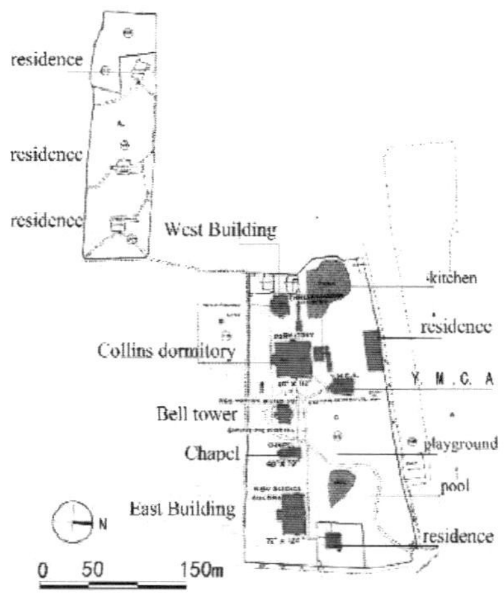

Figure 4 School mapping drawn by Architect & Surveyor design company, New York, USA (1911)

[1]Chen Mingji Construction Factory is the first Chinese factory in Nanjing. It was established in 1888.

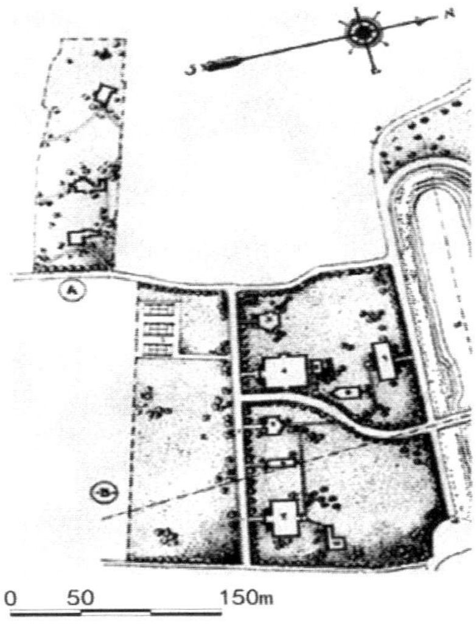

Figure 5 Plan of the United States, New York architect Cody X. Crecory (drawn in 1913)

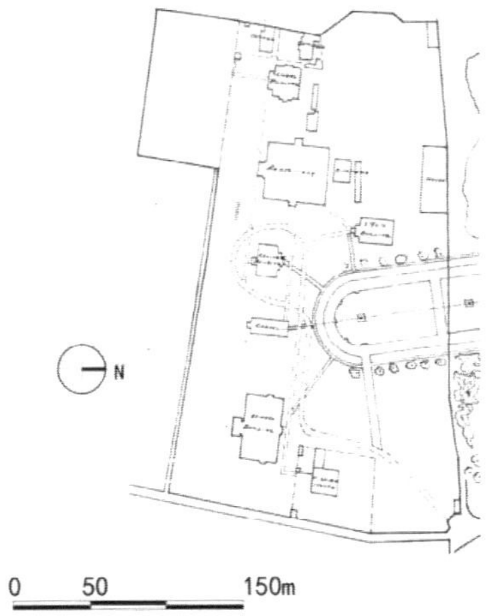

Figure 6 Planning of Jinling Middle School by Perkins, Fellows & Hamilton Architects (drawn in 1914)
Figure 4,5,6 source: Wisconsin Historical Society, USA

Now let's analyze the campus planning and design of Huiwen Academy. Because campus planning and construction of Huiwen Academy were completed at the end of the 19th century and later only part of the roads and environment was adjusted that was not implemented, the original campus planning was used as an example for analysis (Figure 4). In 1911, the Architect & Surveyor Design Company in New York, USA, mapped out the survey of the Huiwen Academy.

IOP Conf. Series: Materials Science and Engineering **809** (2020) 012003 doi:10.1088/1757-899X/809/1/012003

This is the original campus planning of Huiwen Academy (Figure 4). It is learned from the mapping that the campus planning of the Huiwen Academy directly transplants the layout pattern of the "central garden style" of the North American campus. The main buildings are enclosed in the center playground and the green space is concentrated, forming two sets of "central garden style" layout. Form: Bell tower, Chapel, East Building and single teacher residence, Y. M.C. A three-sided enclosure playground; West Building, Collins dormitory, and Y. M.C. A and another single teacher residence on three sides central green space. The main buildings are arranged in a straight line along the main campus road. The Bell tower is the highest point of the campus building. The Bell tower is the center. Chapel is located next to the Bell tower. The East Building, West Building and Collins dormitory are located on both sides of the Bell tower. On the campus, there are two playgrounds in the north and south (Figure 4) .

2.3 Single building of school

There are seven buildings in Huiwen Academy. They are Bell tower, Chapel, West Building, East Building, Collins dormitory,Y. M .C. A.and gymnasium. (Figure 7). Except that the gymnasium was built in 1934, other buildings were built in one time in the late Qing Dynasty (before 1911).

Bell tower Chapel West Building East Building

Collins dormitory Y. M .C. A Gymnasium

Figure 7. The building of the school
source: Nanjing Jinling Middle School History Museum

IOP Conf. Series: Materials Science and Engineering **809** (2020) 012003 doi:10.1088/1757-899X/809/1/012003

The architectural form of all buildings is western architectural form, the external walls are built with blue bricks, and the building structure is made of brick and wood. It's a pity that six of the seven buildings in Huiwen Academy were demolished before 2000. At present, there is only one building - Bell tower.

2.4 Existing building: Bell tower

In 1888, the Bell tower was built. It is the earliest existing building in the Nanjing Church School building. It is now used as an administrative office building and was renovated in 2012. The main body of the building was originally three floors. After the top fire broke out in September 1917, it was repaired by the American Church. The main part was changed to the second floor. The original third floor was changed to the attic, with a tiger window and the original two-slope roof. Changed to a four-slope roof with a cemented diamond tile. The Bell tower was originally five floors, and later changed to four floors. The top bell pavilion has a large clock built in. (Figure 8)

The building's plane is approximately square, 18.2 meters long, 12.4 meters wide and 16.1 meters high. The first floor has a height of 3.9 meters, the second floor has a height of 3.3 meters, the third floor is a loft, and the highest point is 3.2 meters. A tiger window is opened on the sloped roof. The exterior wall of the building is decorated with blue brick feet, the building structure is brick and wood, the brick wall is loaded, and the floor, stairs and roof are all wooden[3]. (Figure 9)

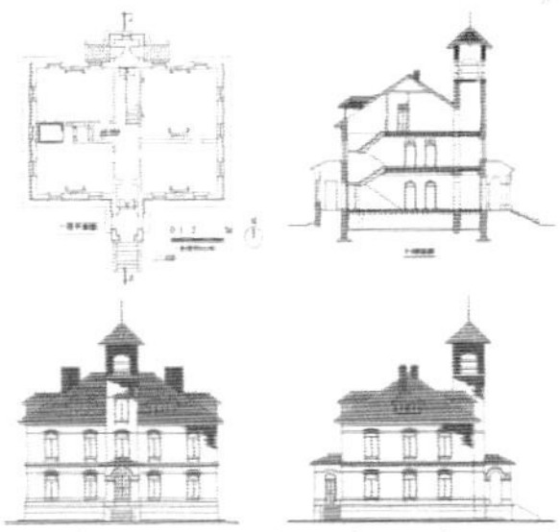

Figure 8 Bell tower (1890), Bell tower (1918), tower source: Yale University Library, USA

Figure 9:The drawing of Bell source: Southeast University, China

3. Reasons for transplanting the Western campus construction model

The relationship between the Christian Church and the school has always been very close. The modern university originated from the medieval university in Europe[2]. Many primary and secondary schools and universities in the American colonial period were founded by the church. The church has rich experience in school construction. So how do you build a campus in Nanjing, China? Western missionaries adopted a campus construction method that directly transplanted the home country of the church, copying the western campus planning and architectural forms.

There are two reasons for this situation: First, after the Opium War, in the face of serious imbalances between Chinese and Western cultures, Western politicians and businessmen have been able to introduce various architectural forms into China without any scruples. They generally have a sense of superiority in Western culture. Second, the Western church schools were arranged in the pattern of traditional Chinese academies. The missionaries did not think that they were particularly suitable. Even if they were adopted, they were the last resort. Chinese classical architecture is a rectangular plane with an aspect ratio of nearly 3:2[3]. The main entrance is located in the middle of the long side of the building. The door is the main activity space of the building. The functional type of the single building is only different with the furnishings. Change is a form of architecture that is constantly changing, and there is no classification concept by type of function. Western architecture has had a single functional type since ancient Greece. It has developed into modern times and the school's functional configuration is complete. Therefore, while introducing the Western education system and teaching content, the church also directly transplants the campus construction model that is suitable for it.

4 Conclusions

At the end of the 19th century, the Huiwen Academy was the largest church school in Nanjing's modern history. While introducing American educational content, the American Church also introduced the American campus construction model. This campus construction model is completely different from the local architecture of Nanjing at that time, providing a model for the

[2] Feng Gang, Lu Bo. Chinese modern university campus under the integration of Chinese and Western cultures [M]. Beijing: Tsinghua University Press: 2016. Page 26

[3] Dong Li. Research on Architectural History of Modern Chinese Christian Universities [M]. Beijing: Science Press, 2010.

construction of schools in Nanjing, which greatly affected the development of modern educational buildings in Nanjing. By clearly combing the development history of church schools in China, this paper has important guiding significance for the heritage protection of today's historical buildings. It also has important reference value for the protection of educational building heritage and campus construction in China and even in the world.

5. References

[1] Zhang Xianwen, 2002. History of the University of Nanking. Nanjing, Nanjing University press.

[2] Yang Zhuwen, 1998. Nanjing Jinling Middle School, Beijing People's Education Press.

[3] Liu Xianjue, Zhang Fuhe, Shin Muramatsu, George Terahara,1992. China Modern Architecture Overview - Nanjing .Beijing, China Building Industry Press.

IOP Conf. Series: Materials Science and Engineering **809** (2020) 012004 doi:10.1088/1757-899X/809/1/012004

Parametric Modeling of Flat Rod Structures

Anna Stefańska[1, a], Wiesław Rokicki[b]

[1]Department of Structure Design, Construction and Technical, Infrastructure, Faculty of Architecture, Warsaw University of Technology, Poland

[a]anna.stefanska2@pw.edu.pl, [b]wieslaw.rokicki@ pw.edu.pl

Abstract. Thanks to the use of advanced digital tools and the achievements of knowledge encompassing numerous fields of science including mathematics, computer science, and even genetics, architects now have the means to create multi-variant design solutions. The quantifiable benefits of interdisciplinary design are increasingly recognized within the field of architecture. As a result, a surprising variety of new ideas and designs emerge. These are often the result of advanced material and structural optimization, and their technological qualities become as important as their visual reception. The subject of this research paper is presenting the results of modeling various configurations of flat rod trusses. The aim of the analysis was comparing the efficiency of the use of materials in differently configured trusses, including trusses created *via* parametric modeling.

1. Introduction

Contemporary architectural design is influenced not only by the developing technologies of construction materials and fabrication processes. The achievements of the first industrial revolution led to the standardization of design elements. Such efforts developed mainly in industrial settings, according to the Fordist idea of standardizing the final product that lead to mass-scale production of unified components. The "think globally, act locally" [1] concept brought on a greater awareness among designers and collaborations in the field of interdisciplinary design. Changes taking place today include "tailor-made"[2] solutions and better opportunities for manufacturing individualized, optimized components. This process can be observed in the increased use of advanced calculation programs. Thanks to such software, it is possible to create mathematical models and algorithms recreating, among others, natural behaviors brought to life *via* biomimetic design processes[3]. The search for optimization of both material and manufacturing technology is also manifest in the selection of efficient structures, materials, etc.

2. Interdisciplinary and topological optimization design

The search for the geometric shape of structures and the construction of rod structures are both important elements in the design of architectural forms, which consists in reflecting the intended visual effect, but also the rationalization of technical and material solutions[4]. From the point of view of tectonics, another important element beyond the shape of the structure is determining the grid distribution[5]. Obtaining effective structural divisions of grids is a task that should be solved as an interdisciplinary issue through analyzing paradigms from two disciplines: architecture and construction. Local compaction and expansion of the grid is an attempt to optimize the structure – too many elements cause the deadweight and then greater deformations. For the most part, the analyses are concerned with determining the metrics of

IOP Conf. Series: Materials Science and Engineering **809** (2020) 012004 doi:10.1088/1757-899X/809/1/012004

the distribution, which in the case of free-form architecture is crucial for the realization of curves describing the form. Large-scale structures, such as bridge and roofing girders, were traditionally designed as relatively straightforward rod structures. The progress in construction took place as a result of changes to geometrical and static models, optimization was usually limited to the use of possibly many repetitive elements. The changes that occur in the manufacturing process of construction elements allow the erection of load-bearing structures from prefabricates that are assembled directly and automatically on the construction site thanks to the use of parametric code(Figure. 1, Figure. 2). The research carried out in this area indicates the possibility of achieving significant efficiency, particularly in regards to the optimization of materials, as evidenced by a significant number of ongoing studies. Thanks to the algorithmization of tools used in the design process it becomes increasingly straightforward to follow the more eco-efficient models derived from research on the evolution of morphogenetic processes occurring in nature. Most of this research, however, is based on material optimization – or the analysis of the flow of forces within structural elements – and the removal of surplus material. When examining grid structures, it is important to find the optimal distribution of rods and nodes so that only those elements that are necessary and do not place additional load on the structure remain in place[6].

Figure 1. An example of generative optimization of material through the use of digital production. Tests conducted by scientists from Imperial College, London in cooperation with ARUP

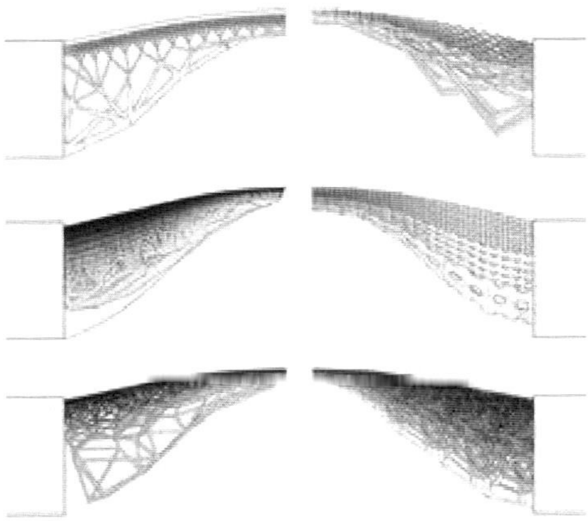

Figure 2. Phases of optimisation of the bridge conducted by scientists from Imperial College, London in cooperation with ARUP

3. Optimization of structural elements

Because they are relatively light, flat trusses are not always adapted to boundary conditions in the optimization process – only their load-bearing capacity is tested[7]. Due to the ease of execution, these constructions are used as girders in bridges, footbridges, or roofs structures. In traditional layouts, it is possible to separate main elements such as top and bottom chords, diagonals, and posts. In majority of projects architects did not attempt to form their structure, instead leaving their visual quality to the engineers. Developed in the 21st century, new software for computer-aided design has amplified the need to alter this way of thinking. The use of generative design methods and the growing awareness of the need to imitate patterns found in nature allow designers to optimize not only the entire shape of the structure, but also the shape of all its components[8]. For example, thanks to Architectural Design Optimization (ADO) analyses, architectural and construction simulations can be conducted, allowing architects to minimize, for example, the use of materials or the energy necessary to produce a support structure. The impact of the development of contemporary analytical methods is particularly evident in the method of shaping structural details. One example of this type of optimization is the research focusing on the elements of connections in membrane structures, conducted in the AGU – the Advanced Geometry Unit, founded in 2000 within the Arup design studio(Figure. 3). Thanks to the current accessibility of additive technologies manufacturing construction elements only on a small scale, it was decided to use FEM to optimize the mass of connectors in one of the objects designed by the studio. The designers faced the task of not only computational material optimization, but also the technological practice of manufacturing the given elements, whose height is the result of applied materials and manufacturing technologies. The final form not only weighed less, but was also significantly different from the original. The optimization allowed saving 75% of the weight of the connectors, resulting in the reduction of the mass of the entire structure by over 40%.

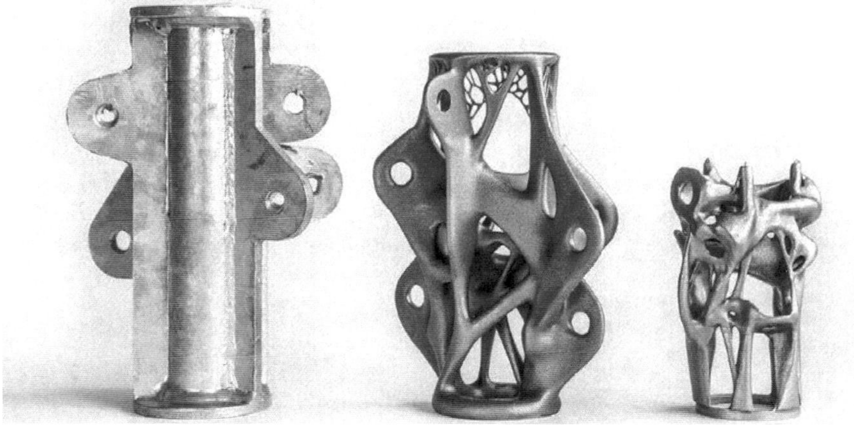

Figure 3. 2 Phases of material optimization of load/weigh, conducted in the interdisciplinary AGU cell within the Arup design studio

4. Development of architectural design optimization (ADO)

A. Gaudi, F. Otto, and S. de Chateau have already carried out optimization activities, however, the inability to calculate geometrically advanced forms has proven to be a serious difficulty. The development of computer-aided design visible at the turn of the 21st century has enabled the introduction of a number of tools that have streamlined computer-aided calculations. In a way, the successive generations of new software force users to search for new aspects of optimization. Presently, it is possible to use advanced mathematical algorithms and improve the technical qualities of materials, including in the nanoscale. The surrounding natural environment affects man's creative intuitiveness – it is a source of inspiration, a guide. Contemporary shaping of structural forms of architecture consciously draws on bionic patterns. These actions appear to be dictated, on the one hand, by the search for a new visual quality in architecture, and on the other – the rationalization of technical and material solutions. One should also remember that the structure is subject to the laws of physics, construction and building materials age with the passage of time, while the form and function are adapted to the environment, as well as assessed on the basis of subjective impressions – and all this stems from Nature(Tarczewski 2011). Digitized design's capabilities for creating new architectural forms depend on the perceptive and cognitive abilities of architects. Meanwhile, the parametric and generative design process is a dynamic process characterized by discovering new possibilities in comparing numerous variants. In architectural searches, particular attention is paid to form finding and topology.

The role of the designer in using generative tools is to interpret countless results by creating parametric models or genetic algorithms, analyzed in static calculations while creating topologically variable surfaces, isomorphic skeletons, the emergence of new forms, and self-similarity processes, etc. The significant geometric complexity of structural forms is a big challenge in contemporary architecture. More and more often, unique objects in innovative shapes drawn from nature are being implemented. Having gained access to ever more powerful computer-based calculation capabilities and improved methods and methods of fabrication, architects are able to design increasingly sophisticated objects.

5.In the search of optimal rod structures-case studies

The search for effective engineering solutions in keeping with architectural values prompted the authors to analyze non-standard flat trusses. The research presents various ways of topological optimization. Obtaining the minimal mass in each of the considered variants was adopted as the most important parameter. Differences in the number of nodes and the length of rods have been disregarded due to the fact that flat rod structures are fairly simple elements to fabricate. Conclusions from the research conducted in the context of design of load-bearing structures with some degree of geometric complexity can become a useful tool in the hands of an architect. The scripts prepared during the research were designed to universally adapt to boundary conditions, depending on the adopted configuration of rods. The assumed task was to develop basic assumptions and preliminary guidelines for the design of trusses with a free-form rod structure. Different geometries were a part of authors case studies Figure 4. presents standardized types of traditional trusses, Figure 5. presents configurations of parametric designed trusses.

The longest truss has a smaller mass with a height/span ratio of 1/11L. These small differences, however, are insignificant and therefore are disregarded in further analysis. As the next step, the profile cross-section was optimized using Robot Structural Analysis software: 4 groups of rods were generated for each solution. Due to dividing rods into groups and the assumption of different closed cross-sections, each of the solutions achieved a significant reduction in the mass of the truss (between 9% and 50%).

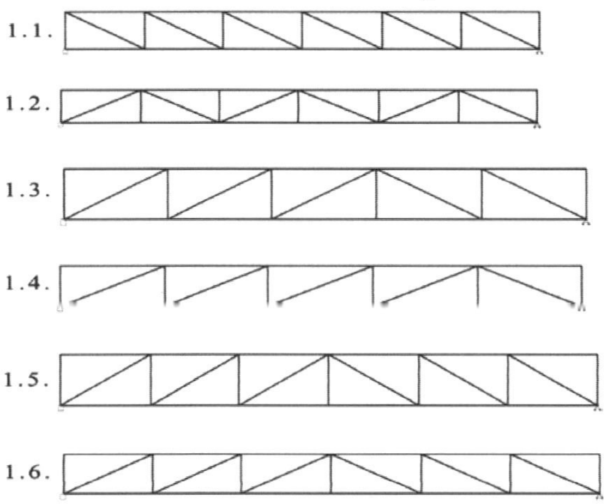

Figure 4. Case studies
- shape of models in Analysis 1

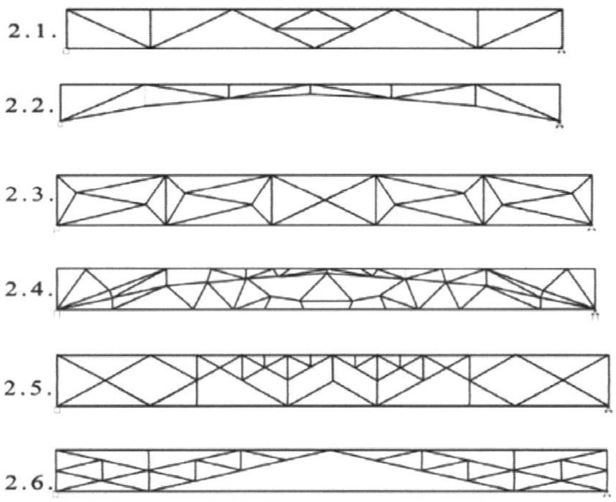

Figure 5. Case studies
- shape of models in Analysis 2

Table 1. Analysis 1 – Summary of truss patterns and total combined mass for each individual variant

#	Truss/ Lattice construction structure length [m]	Truss height Proportions h = 1/11L or h +1/14L	Weight [kg] for 1 group of rods	Weight [kg] for multiple groups of rods	Optimization effect resulting from changing the profiles of rods
1.1.	13,5	1.22	351	319	-9,11%
1.2.	13,5	0,96	348	174	-50,00%
1.3.	16,25	1,47	400	263	-34,25%
1.4.	16,25	1,16	386	243	-37,04%
1.5.	16,5	1,5	421	323	-23,28%
1.6.	16,5	1,17	460	316	-31,30%

The modifications undertaken in Analysis 2 consisted in changing the internal structure of the trusses, while maintaining the location of the supports, the external shape and the point of application of external forces (the requirement to keep these truss nodes in these positions). The adopted variants are free-form interpretations of the geometric truss developed on the basis of an algorithm. The results described already include the material optimization described in Analysis 1. Variants 2.1 and 2.5 show a significant reduction in the mass of the entire structure – 37,93% and 28,79% respectively – when compared to the optimized variants from Analysis 1. This results from the distribution of truss nodes and rods in relation to the distribution of forces. In addition, example 2.1 does not have a different geometry than standard trusses, which suggests that the optimization process does not require excessive increase in the total length of rods or the total number of nodes in order to achieve satisfactory results. In example 2.2, despite a significant transformation of the geometry and the reduction of the total length of rods, the mass of the truss did not change. In example 2.4, despite a slight increase in weight, the tendency of the nodes adapting to the forces involved can be observed. In the next steps, in order to optimize, it would be necessary to eliminate rods within the bottom chord, which add extra load to the structure. Variants 2.3 and 2.6, which were generated in the process of symmetrical truss divisions, are not designed in accordance with the flow of forces, and therefore the entire structure must bear the load of the excessive number of ineffective rods.

In traditional truss design(Table 1), one of the most important parameters is to determine the height of the truss in relation to its span. However, as it can be seen in the analyses, the search for a more favorable rod structure in the construction of the truss is also important. Shaping the form with the use of generative design tools allows obtaining many solutions that can determine the quality of the architectural message(Table 2). Striving for optimization of, among others, the minimal use of material, does not limit architects to the design of classic geometric forms.

Table 2. Analysis 2 – Summary of truss patterns and total combined mass for each individual variant

#	Truss/ Lattice construction structure length [m]	Truss height Proportions h = 1/11L or h +1/14L	Weight [kg] for multiple groups of rods	Optimization effect resulting from changing the profiles of rods
2.1.	13,5	1.22	198	-37,93%
2.2.	13,5	0,96	173	0,00%
2.3.	16,25	1,47	378	+43,72%
2.4.	16,25	1,16	255	+4,93%
2.5.	16,5	1,5	230	-28,79%
2.6.	16,5	1,17	398	+25,94%

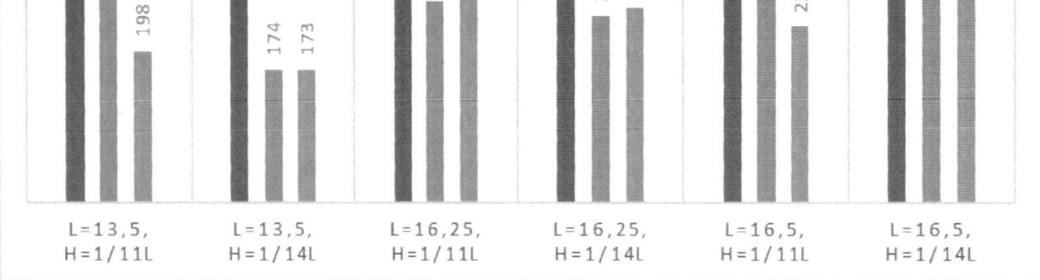

Figure 6. Comparison of all variants

IOP Conf. Series: Materials Science and Engineering **809** (2020) 012004 doi:10.1088/1757-899X/809/1/012004

6. Conclusions

Reflections on how to optimize rod structures have been developing since the 19th century, but it was access to contemporary computer-aided programs has advanced these efforts. Differentiation, but also the creation of free form and geometrically complex structures is a feature of contemporary architecture. The original, bionic shapes of buildings erected in the 21st century are the result of explorations in avant-garde architecture – surprising in terms of art, and increasingly rational due to the technical and material solutions they employ. Computational calculation methods lead to the improvement of modeling tools, as well as the fabrication of construction elements, and architects are able to design unconventional forms with increasingly complex structures. The algorithmization of tools supporting the design process provides new opportunities in the search for innovative and effective load-bearing structures. The results presented above indicate that there are many possibilities in the search for optimized material solutions in architectural and construction design, which is particularly important in the context of more economical management of resources available on the global scale.

Acknowledgement
The paper was carried out as a part of grant at Warsaw University of Technology, Faculty of Architecture.

7. References

[1] Moberg C. L. Rene Dubs, Friend of the Good Earth: Microbiologist, Medical Science, Environmentalst, Washington, DC,: ASM Press, 2005

[2] Januszkiewicz K., *Projektowanie parametryczne oraz parametryczne narzędzia cyfrowe w projektowaniu architektonicznym.* Parametric design and parametric digital tool in architectural design, Architectuae et Arbitus 8, Bialystok, 2016

[3] Tarczewski R., *Topologia form strukturalnych; naturalne i tworzone przez czlowieka prototypy* [Topology of structural forms; natural and manmade prototypes of structural forms in architecture], Publishing House Oficyna Wydawnicza Politechniki Wrocławskiej, Wrocław 2011, p.13

[4] Rokicki W., *Structure in aeurythmic architecture,* Konstrukcja w aeurytmicznej architekturze, Oficyna Wydawnicza Politechniki Warszawskiej, Warsaw, 2006

[5] Gawell E., Stefańska A., *The idea of fabrication of structural elements in shaping modern pavilions*, „Zeszyty Naukowe Vistula", Warszawa, 2018, pp.82-92

[6] Nervi, P. L. *Concrete and structural form*, The Architect and Building News, 208(27), 1955, pp.523-529

[7] Gaynor A., Guest J., Moen C., *Reinforced Concrete Force Visualization and Design Using Bilinear Truss-Continuum Topology Optimization*, American Society of Civil Engineers, 2013

[8] Sigmund O., *Morphology-based black and white filters for topology optimization*, 2006

Research on Adaptive Facade Renovation Design of Industrial Architectural Heritage: A Case Study of Nanjing Hutchison Factory

Yanming Wu[1, a], **Qi Zhou**[2, b]

[1] School of Architecture, Southeast University, Nanjing, China
2 School of Architecture, Southeast University, Nanjing, China

[a]791367448@qq.com, [b]zhouqiseu@126.com

Abstract. The pst-industrial era witnesses the gradual decline or transformation of China's traditional industrial industry. Meanwhile, the original facade design of most of the existing industrial architectural heritage can no longer keep up with the pace of the times, which left a lot of issues that need to be considered and studied urgently in the renovation of industrial architectural heritage.
In this article, the adaptive facade renovation design of industrial architectural heritage was studied and effective design strategies were summarized. Firstly, the research background of adaptive renovation of industrial architectural heritage and the research objects in this article were outlined and discussed, and then the relevant concepts and materials of facade renovation of industrial architectural heritage were summarized. Secondly, the renovation of Nanjing Hutchison Factory was cited here as an example for specific analysis. Finally, based on the analysis and summary of the case, corresponding design strategies and methods for the adaptive facade renovation were summarized, and the future research direction was put forward. This article is not only of great guidance and promotion to the renovation of Hutchison, but also of certain reference value to similar facade renovation projects of architectural heritages in China and even in the world.

1. Introduction

Entering the 21st century, the Chinese government strongly advocates sustainable development under the joint force driven by energy crisis, environmental protection, resource recycling and urban cultural construction, etc... Construction has seen large-scale demolition, reconstruction and renovation. The original facade design of most of the existing industrial architectural heritage can no longer keep up with the pace of the times, not to mention shaping a good image of the city and creating a unique artistic conception of the city. Therefore, the adaptive renovation of the facade of the industrial architectural heritage has become a key content in the current construction and renovation work. With the development of new building materials, new technologies and diversified building theories, the facade renovation of industrial architectural heritage is no longer limited to simple traditional methods, but is performed by using similar materials to perform facade reconstruction while keeping the historical context.

The research scope of this article is limited to the adaptive facade renovation of Nanjing Hutchison Factory along the Yangtze River in Nanjing, aiming at summarizing the corresponding design strategies for facade renovation based on the case study of Nanjing Hutchison Factory, an industrial heritage complex. Nanjing Hutchison Factory is the largest existing industrial heritage complex in modern China. Located at 168 Baotaqiao West Street, Xiaguan District, Nanjing City, Jiangsu Province, it was founded in 1912 as the first foreign-funded food processing factory in Nanjing. Nanjing Hutchison Factory was designed by Norwegian architect E. J. Muller and constructed by Yao Xin Construction Factory. In September 2015, the original site of Nanjing Hutchison Factory was basically demolished, but there were still more than 10 historical buildings built in 1920-1930, including office building, laboratory building, factory building, slaughter room and cold storage, with a total construction area of about 70,000 square meters. As one of the modern factories built in China by its investors William Vestey and Edmund Vestey, Nanjing Hutchison Factory is a model of the combination of western modern technology and production technology with local technology and craftsmanship, and is also a sample of the integration of architectural style and technology (refer with: Figure. 1). Therefore, it is of great significance to select such an industrial architectural heritage as the research object in this article. The author had the honor to participate in the renovation design of Nanjing Hutchison Factory as one of the assistants of the project manager in the autumn of 2017. Here, the concept, method and corresponding strategy of the adaptive renovation of the facade of industrial architectural heritage will be discussed based on the author's experience in the whole project design.

Figure 1. Aerial View of Nanjing Hutchison Factory

The focus is put on the facade renovation design of industrial architectural heritage. In Chapter Two, the concept and related materials of adaptive facade renovation are summarized, and the general principles and commonly used facade materials of adaptive facade renovation are explored and analyzed. In Chapter Three, the principles followed in the renovation design, the facade materials used and the corresponding effect feedback of Nanjing Hutchison Factory are illustrated. In Chapter Four, three design strategies for the adaptive facade renovation of industrial architectural heritage are proposed. Finally, the significance of the adaptive facade renovation of industrial architectural heritage is concluded.

2. Concept and related Materials of Adaptive Facade Renovation of Industrial Architectural Heritage

It is necessary to pay attention to the general principles affecting the design in the adaptive facade renovation design of industrial building heritage, so that the completed building can meet the basic requirements of all aspects. It should be said that any building facade design has principles to follow, but different building design projects attach varied importance to the principles and show different emphases or understandings.

Four general principles, namely, the principle of coordination between function and individuality, the principle of visual aesthetics, the principle of durability and easy maintenance, and the principle of environmental protection and sustainability, were put forward for adaptive facade renovation design. Specifically speaking, first of all, the renovation of the facade should conform to the principle of coordination between function and individuality, which means that the renovated works shall be able to meet the requirements on function and use. It can neither be done simply nor pursued deliberately. We should respect the industrial architectural heritage, ensure that the individuality of materials is consistent with the functions of buildings, and use rigorous analysis methods to inspire our own creative inspiration and design architectural works that meet the actual functional needs.[1] If the renovation is carried out roughly without considering the coordination between the architectural function and the material individuality, it will cause confusion to people, which is thus an irresponsible behavior. Secondly, the renovation of the facade should conform to the principle of visual aesthetics, including people's basic feeling of visual elements such as shape, light, color and space of the architectural appearance. Material performance shall conform to the architectural aesthetic standards under the influence of current society, consumer culture and popular culture, and facade renovation design shall be carried out according to appropriate proportion, color and texture. Only by incorporating new design techniques or giving it a brand-new appearance on the original noumenon framework, and removing the dross from its essence, can the lingering charm of industrial architectural heritage be preserved while maintaining the overall aesthetic feeling of the industrial architectural heritage. Thirdly, the renovation of the facade should conform to the principle of durability and easy maintenance. We should reduce unnecessary costs and waste, make old buildings look brand-new on the premise of satisfying practical functions, reduce unnecessary economic expenditures, and save resources and create economic value. With the passage of time, the materials of the exterior walls of the industrial architectural heritage will be rotten and dilapidated, and we should follow the principle of practicality, beauty and economy to renovate the historical buildings. Finally, the renovation of the facade should conform to the principle of environmental protection and sustainability. Nowadays, in order to meet the design requirements, designers have gradually abandoned natural materials and adopted a large number of industrial materials that are not environmentally friendly or have short service life, thus causing more and more environmental problems. We are beginning to realize that instead of relying on natural materials, we have ironically lost the foundation on which we maintain our current lifestyle.[2] As early as 1998, Mr. Wu Liangyong put forward in "Prospect of Architecture in 21st Century " that "in an era of coexistence of multiple technologies, high and new technologies, appropriate technologies and traditional technologies should be developed jointly according to local conditions. We need to see this direction and clearly recognize that appropriate technologies themselves have simple sustainable development guidelines as times change."[3] Therefore, in the process of facade renovation, we must follow the principles of environmental protection and sustainability and reasonably select materials. From this, we can see that the four general principles mentioned above contain rich project design experience and philosophical thoughts, which should also be of great reference value and guidance to the renovation design of Nanjing Hutchison Factory.

However, since the exterior wall of the buildings is constructed from various materials, no matter how perfect the principle is, the facade renovation design cannot be completed without in-depth study of the materials. The purpose of material performance is to make use of the characteristics of different materials in shape, color and texture, and through careful organization and arrangement by designers, to produce beautiful architectural forms. So, different building materials can bring a completely different form of buildings, and this beautiful architectural form can be perfectly preserved in the renovation of old buildings.[4] Therefore, analyzing the materials, understanding the properties of materials and exploring the performance characteristics of materials (such as shape, color, texture and texture, regional characteristic, etc.) can play a favorable guiding role in the design and implementation of specific building projects.

Exterior wall materials used in the renovation of China's industrial architectural heritage mainly include *lacquer*, *cement mortar*, *washed granolithic plaster* and *stone granular facing mortar*, etc. (refer with: Figure. 2). *Lacquer* is a kind of paint imitating marble and granite. It is mainly applied to imitate the stone effect of exterior walls. It has the characteristics of fire prevention, waterproof, acid and alkali resistance, pollution resistance, non-toxic, tasteless, strong adhesion and never fading. Replacing dry hanging of stone with lacquer not only reduces the cost, but also reduces potential safety hazards.

Moreover, the layout can be designed and planned at will, which is convenient and fast. It accelerates the construction progress and improves the overall aesthetic appearance of the building. *Cement mortar* is made of mortar, cement, gravel and sand according to a certain mix proportion and prepared manually according to the original process. *Washed granolithic plaster* represents a traditional construction technology, which can make the texture of the wall surface natural, the color solemn and beautiful, and the decorative surface firm and durable. It does not fade, and has better resistance to pollution. *Stone granular facing mortar* is an overcoating material composed of natural stone particles with different particle sizes. It can be used for colorful decoration, with soft color and good durability. As common building materials for restoring modern architectural heritage in China, these materials have been enduring in the development process of China's architecture and are favored by the majority of architects. Of course, with the progress of technology, there are other materials that have also played an important role in the facade renovation of the architectural heritage. Due to the limited space of this article, they will not be discussed here for the time being.

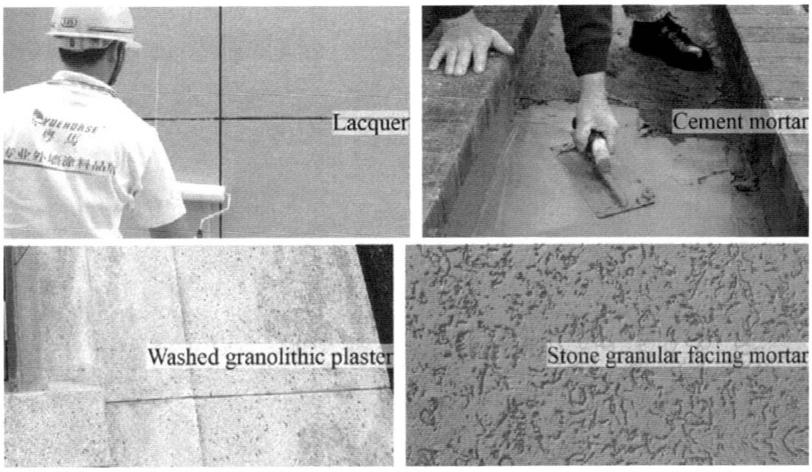

Figure 2. Four major exterior wall materials used in the renovation of china's industrial architectural heritage

As Nanjing Hutchison Factory is located along the Nanjing Yangtze River Bridge, it has advantageous geographical location and profound historical and cultural background. Hence, the renovation design of it has naturally become the top priority of Nanjing and even the whole country. As the gateway to Nanjing, the image of its complex is particularly important. Therefore, the selection of appropriate design principles and materials for facade renovation has become the top priority in the design of this project.

3. Principles and Materials for Facade Renovation Design of Nanjing Hutchison Factory

Nanjing Hutchison Factory used to be a food processing factory invested by foreign capital in history. Because of the renewal of the city, its actual function has declined. In addition, due to the long history, the external wall of the original buildings is seriously damaged, and the architectural style cannot fit the existing city style, but its architectural space still has use value. Therefore, the Nanjing Municipal People's Government decided to redesign such an industrial heritage complex to maximize the vitality of the area. As a designer, our team made full use of and developed the general principles discussed above in renovation design of Nanjing Hutchison Factory, as detailed below:

3.1. Principle of Coordination between Function and Individuality.

Under normal circumstances, window opening is not required in renovation of the general facade of commercial buildings, and the application of materials is more flexible. Metal wrapping and glass curtain wall are mostly used for material reconstruction. If the renovated industrial building heritage is used for office or residence, lighting and other factors need to be considered. At this time, the application of materials should be based on the environment in which the building is located, and should not be exaggerated. The principle of coordination between function and individuality should be followed in such circumstances. In the facade renovation design of Nanjing Hutchison Factory, the design team analyzed the features and values of the facades of various historical buildings in advance and then carried out appropriate renovation design on the basis of fully retaining the historical marks so as to ensure that the renovated building facade can meet the needs of internal functions. For example, cinemas and auditoriums have lower requirements for daylighting, so they were mainly arranged in closed exterior walls. Commerce and repast have higher requirements for daylighting, so they were mainly arranged in exterior walls with large windows (refer with: Figure. 3).

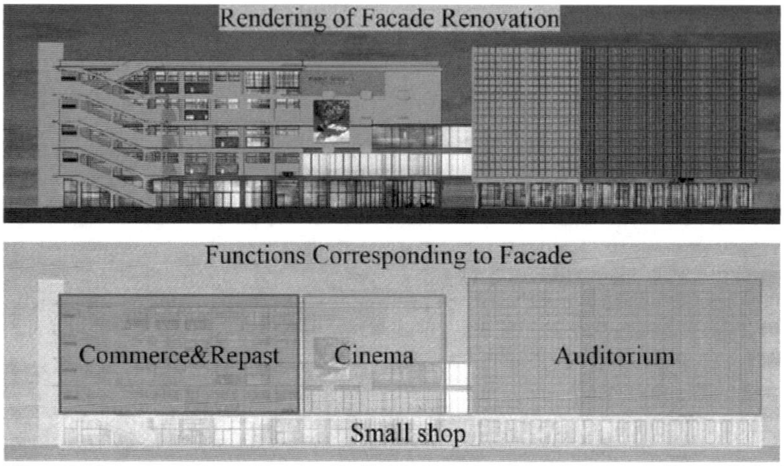

Figure 3. Coordination between Function and Individuality in Facade Renovation of Nanjing Hutchison Factory

3.2. Principle of Visual Aesthetics.
Generally speaking, the facade design of historical buildings will follow a set of classical rules of order and the rigorous geometric order will derive its inherent beauty of order. Therefore, Nanjing Hutchison Factory attaches great importance to the regeneration of old buildings in the process of design, takes its essence and discards its dross, preserves the sense of order, and maintains the geometrically classical beauty of order through the integration of new design techniques (refer with: Figure. 4).

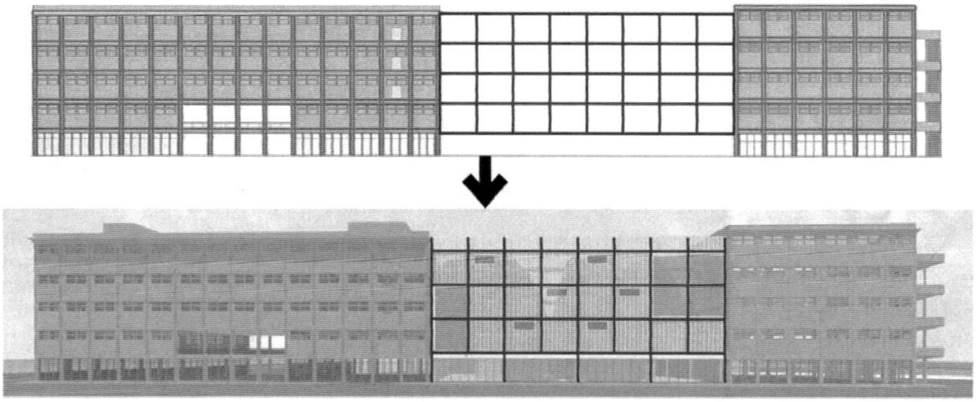

Figure 4. The Facade Restructuring of Nanjing Hutchison Factory and Its Application to New Buildings (photo source: by the author)

3.3. Principle of Durability and Easy Maintenance.

With the passage of time, the materials of the exterior walls of Nanjing Hutchison Factory were rotten and dilapidated, and we followed the principle of practicality, beauty and economy to renovate the historical buildings. In the selection and application of materials, we considered the durability and easy maintenance of the materials to reduce unnecessary cost and waste and make the old buildings look brand-new.

The exterior wall facing design of Nanjing Hutchison Factory fully reflects the texture of the original materials. When we designed, we adopted cement mortar facing just as that used in traditional buildings. Through mixing different materials and modulating colors, the exterior wall materials of historical buildings are re-imitated to make the historical buildings glow with vitality.

4. Facade Renovation Design Strategy of Nanjing Hutchison Factory

Although the facade renovation of the complex of Nanjing Hutchison Factory followed the appropriate principle and used the appropriate materials in the process of facade renovation design, our design team still faced many details that ran counter to the expected results in the process of project construction. For example, if the facade painting method was not guided by the correct design strategy, it is likely to produce different facade effects (refer with: Figure. 5). Therefore, it is particularly important to further summarize the relevant design strategies for adaptive facade renovation. According to the performance characteristics of the facade materials of industrial architectural heritage, the article attempts to divide the facade renovation strategy into four main aspects: Harmony of color strategy, material shape and scale reconstruction strategy, material texture analogy strategy, and material fragment retention strategy, which are further explained and discussed below.

Figure. 5 Comparison of Test Samples Using Stone Granular Facing Mortar for Exterior Wall Repair

4.1. Harmony of Color Strategy- Harmony of Color Technique.

The color of materials can become a "binder" for connecting various elements on the building facade, and also a "repellent" for separating various elements.[5] When the color of the facade materials of industrial buildings tends to be consistent in the proportion and brightness, the whole will present a coordinated and integrated effect. Therefore, architects can create mutually harmonized colors on the original facade materials to achieve an overall harmonious effect, and this is called the Harmony of Color Technique.

In the process of reconstructing the facade materials of industrial architectural heritage, it is inevitable to encounter the problem of conflict between the old and new materials, that is, the problem of inconsistent material language. At this time, the Harmony of Color Technique is an effective solution. This technique can not only present the charm of the exterior wall material of industrial architectural heritage in a brand new manner, but also fully solve the problem of the fusion of new material color and the original color. In this case, it acts as a "binder". In the process of restructuring the exterior wall materials of historical

IOP Conf. Series: Materials Science and Engineering **809** (2020) 012005 doi:10.1088/1757-899X/809/1/012005

buildings of Nanjing Hutchison Factory, we imitated the interface colors of the industrial architectural heritage according to its characteristics and the surrounding environment, highlighting the sense of the times and organically integrating the old and new styles (refer with: Figure. 6). The Harmony of Color Technique makes the facade renovation of Nanjing Hutchison Factory well blended with the surrounding environment. And the fusion of slight difference in tone and brightness between materials can effectively make buildings well integrated into the surrounding environment. It helps form a new coordination relationship without abruptness.

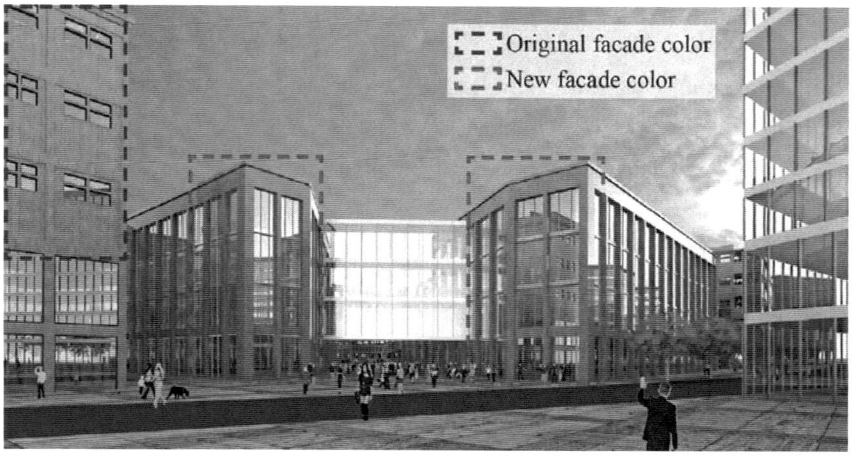

Figure. 6 Adaptive Renovation Design of the Color of the Exterior Walls of Nanjing Hutchison Factory

4.2. Material Shape and Scale Reconstruction Strategy.

4.2.1. Grading of Material Shape and Scale

Architecture has the classification of shape and scale. Similarly, materials also have the classification of shape and scale. In the adaptive renovation design of facade material shape of industrial architectural heritage, the design focus is to integrate shape and scale for materials of different scales in different periods and coordinate the proportion between old and new elements.

T. Hamlin, an American modern architect, thinks that "In architecture, scale makes the building present a certain proper size as expected, which is a unique characteristic that seems to be inherent in architecture."[6] According to this, this kind of proportion relation should meet people's psychological and visual requirements. Different proportions of materials give people different feelings. Designers express various understandings of buildings through the treatment of different shapes and scales of materials.

As for the material of cement, its shape and scale can be cut in different dimensions to form prefabricated modules by technical means as required. In the renovation of facade materials of Nanjing Hutchison Factory, we classified the old and new materials according to their length, area, volume, specific structural details and decorative details to eliminate the obvious difference between the old and new materials. The transition of scale is formed between materials, which gives people a sense of coordination and makes the scale expression clear and distinguishable. At the same time, the reserved long glass window forms a strong contrast to the small brick material, thus presenting a sense of rhythm (refer with: Figure. 7). This method is widely used in the renovation of industrial architectural heritage.

IOP Conf. Series: Materials Science and Engineering **809** (2020) 012005 doi:10.1088/1757-899X/809/1/012005

Figure. 7 Adaptive Renovation Design of the Cement Module of the Office beside Nanjing Hutchison Factory

4.2.2. Reorganization of Material Shape and Scale- Modulus Analogy Technique

In the facade renovation of industrial architectural heritage, there is a kind of logical relationship between the new material and the old material in terms of shape and scale, which makes the new material and old material match quite well after renovation. And this is called modulus analogy technique. In applying this technique, we should pay attention to the following points. On one hand, the shape of the original materials should be strictly distinguished to prevent confusion; on the other hand, the shape and scale of the new materials should be correlated with the old ones so as to achieve the effect of synergy.

In facade renovation of Nanjing Hutchison Factory, the glass curtain wall materials of the new building adopted modulus analogy technique, i.e. the glass curtain wall was divided into many groups on the X axis, and the width of each group corresponds to the width of the opening of the facade of the historical architecture, thus echoing the historical structural characteristics of the old building and completing the adaptive renovation design for shape and scale (refer with: Figure. 8). The facade surface pieced together by different combinations can also arouse people's remembrance of the old factory building.

Figure 8. Adaptive Renovation Design of The Glass Curtain Wall Materials of Nanjing Hutchison Factory

4.3. Material Texture Analogy Strategy.

Texture of materials plays an important role in material performance. In renovation of existing buildings, the introduction of new materials is bound to produce differences in texture. Therefore, how to make the old and new materials well blended in texture requires advanced technology and philosophy.

In decomposition of the texture of the existing buildings facade, it is necessary to preserve, extract and summarize the historical symbols, interpret the historical information of the materials, and awaken and heighten such historical symbols and information in renovation. In doing so, the historical sense of the buildings can be preserved and a new image will be presented, thus bringing a collision between history and modernity. In adaptive renovation design of the historical buildings of Nanjing Hutchison Factory, stone granular facing mortar was used to restore the charm of the historic buildings. In this process, in order to make the restored facade of the historical buildings perfectly integrate into the well-preserved facade of the existing historical buildings, the texture of the exterior wall materials for the new buildings should be as close as possible to the facade texture of the historical buildings preserved on site. Therefore, how to allocate the proportion of sand, stones and other basic materials is particularly important (refer with: Figure. 9).

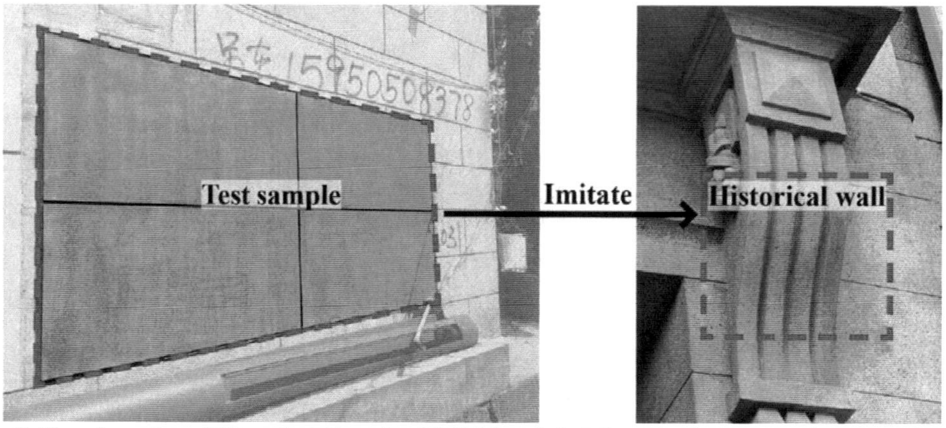

Figure 9. Test Samples of Historical Exterior Wall Materials Mixed in the Renovation Site of Nanjing Hutchison Factory

4.4. Material Fragment Retention Strategy.

"The feeling of detail is an indispensable part in aesthetics. It is fundamental to the basic aesthetic choice and the complex reflection of comments. It is 'rooted' in feelings."[7] For those buildings with high retained value and rich historical information, the color, shape and texture of the original facade materials of industrial architectural heritage will not be re-designed for adaptive renovation, but the material fragments will be selectively reserved or some detail components will be added, so as to enrich the form of industrial architectural heritage and strengthen the existence and identifiability of buildings. The relationship between material details and architecture is complementary. The choice of detail decoration depends on the volume of architecture. Generally speaking, the adaptive renovation design of material details should be relatively simple and distinguished, while buildings of small volumes should choose delicate and exquisite details.

Figure. 10 shows a reconstruction of the "northern factory building" of Nanjing Hutchison Factory. The exterior wall of this building has been dilapidated before. In order to maintain the characteristics of historical buildings and to maintain the visual continuity of the building facade, the designers used new steel to reinforce the original wall structure, replaced the incomplete windows of historical buildings with new steel windows, and retained the original water tower on the roof. The preserved water tower looks more exquisite and touching on the restored building ontology. Elements that seem to have failed to adapt to the development of the times can still be stimulated to achieve rational modeling effect as long as they are treated prudently and put in a correct position. The water tower placed in the roof garden seems to have a direct dialogue with the tourists. It records the passage of time, arouses people's emotion and collective memory of the place, and continues the breath of the old factory.

Figure 10. Adaptive Renovation Design Strategy of Fragmental Retention of Nanjing Hutchison Factory

5. Summary

Four design strategies for facade renovation of industrial architectural heritage were proposed based on the exploration of the general principles and materials applied in adaptive facade renovation of industrial architectural heritage with Nanjing Hutchison Factory as the study case. The research is not only of great significance to guide and promote the renovation of Nanjing Hutchison Factory and preserve its historical essence, but also of certain reference value to similar facade renovation projects of architectural heritages in China and even in the world.

China has left over a large number of abandoned buildings in its rapid development, which reflects the complexity of the problem. Limited by culture, regional materials and research objects, the design strategies summarized in this article cannot solve all the problems of facade renovation of existing buildings in the world. Therefore, the author plans to make in-depth investigation of existing architectural renovation cases in more regions, so as to go further in the research on facade renovation of architectural heritage.

Acknowledgement

This work was supported by the *National Natural Science Foundation of China* (No. 51778123). From 2015 to September 2016, the renovation project of Nanjing Hutchison Factory was designed by Zhou Qi Studio, and from September 2016, the renovation project was jointly undertaken by Zhou Qi Studio and Benoy Design Company. I had the honor to join Zhou Qi Studio in September 2017 and participated in renovation project of Nanjing Hutchison Factory as an assistant designer. The design team's scheme pictures quoted in the article have all been illustrated.

6. References

[1] Zhang. W, Chen. J, & Liang. X: "Analysis of External Form Reconstruction of Old Buildings," New Arch,1, 90-93, 2006.
[2] Bergman. D: Sustainable Design Essentials Guide (X. Xu & R. Chen, Trans.). Jiangsu Science and Technology Press, 2014.
[3] Wu. L: "Prospects of Architecture in the 21st Century," Urban Planning,6, 10-21, 1998.
[4] Zhu. S, & Wang. L: "Intervention and Application of Modern Materials in Old Buildings," Science and Technology Innovation and Application,15, 238-238, 2012.
[5] Francis. D. K: Architecture: Form, Space and Order. (C. & D. Zou, Trans.). China Building Industry Press. 1998.
[6] Hamlin. T, & Zou. D: "The principle of architectural form beauty," in Architecture and Culture,2, Beijing: China Building Industry Press. pp. 98-98. 1990.
[7] Scruden. R: Architectural Aesthetics (Vol. 102) (X. Liu, Trans.). Jianding Publishing. 2005.

IOP Conf. Series: Materials Science and Engineering **809** (2020) 012006 doi:10.1088/1757-899X/809/1/012006

Wind Loads of an Irregularly Shaped Tall Building in Warsaw – Eurocode 1 Standard Analysis and Wind Tunnel Testing

Joanna Pietrzak[1, a], Mariusz Wrona[1] and Wiesław Rokicki[1]

[1]Department of Structure Design, Construction and Technical Infrastructure, Faculty of Architecture, Warsaw University of Technology, Warsaw, Poland

[a]email: joanna.pietrzak@pw.edu.pl

Abstract. This paper analyses wind loads of an irregularly shaped skyscraper in Warsaw. The aim of this study was to compare the results of two methods: analysis conducted according to the Eurocode 1 Standard and wind tunnel testing. The wind pressure distribution on façades and net force values were analysed. In the Eurocode-based analysis, the role played by the form of the building is treated in general terms and the aerodynamic interference with surrounding buildings is not taken into account. The results are characterized by a high design factor of safety. The results of wind tunnel experiments are much more accurate, more reliable, and could be the basis for bassuming lower wind loads.

1. Introduction

The construction of modern tall buildings requires solving increasingly complicated architectural and construction problems, structural, installation, and communication issues, as well as technical and fire safety problems. It should be noted that the scale of difficulty in the design process and implementation technologies is primarily related to the height of the buildings. With regard to the load-bearing structure, a building is considered as a tall building when the process of designing its load-bearing structure is decisively affected by wind load [1]. However, the impact of wind on a given structure is strongly influenced not only by the latter's height but also its shape and other buildings in its immediate surroundings. Determining the location of a skyscraper, especially within dense urban tissue, requires relatively precise aerodynamic [2].

When designing a high-rise building, one should strive to limit excessive tension in structural elements [3], bearing in mind that the spatial form greatly affects the size and distribution of forces [4]. The interaction between the form and wind determines the air mass flows around the building and wind pressure distributions on the façades [2]. To obtain more effective solutions, it is important to perform aerodynamic optimization analyses at the initial stage of the development of architectural concept. The possibility to eliminate the adverse impact of wind in the context of obtaining an innovative spatial form, as well as the possible reduction of construction costs, results in a greater interest in wind engineering research on the part of investors and designers [5].

In order to optimize the form and construction of a skyscraper, increasingly precise tools are needed to better understand the size and nature of aerodynamic interactions. This applies, in particular, to the erection of irregularly shaped buildings whose complexity means that the ability to accurately recognize aerodynamic phenomena is related to the chosen method of aerodynamic analysis. In Europe, wind tunnel testing is recommended for buildings higher than 200 metres. When designing buildings lower than 200 meters (which constitute over 96 per cent of European high-rise buildings), loads arising from the effects of wind can also be assumed according to *Eurocode 1 European Standard* [6, 7]. This essay presents the results of a comparative analysis carried out to verify the loads assumed in accordance with *Eurocode 1* in relation to the results of wind tunnel experiments.

2. Subject of the study

Despite the fact that simple, geometric shapes characterize most of the high-rise buildings currently being erected in Europe, there is a visible, emerging tendency to design irregular forms. It is particularly necessary to conduct more precise analyses of spatial forms that cause the formation of adverse aerodynamic phenomena, for example: non-aerodynamic, asymmetrical, irregular forms, or those positioned transversely to the wind direction [8]. Objects located in urban space, in an environment that causes difficult to predict wind effects also require detailed analysis. This can be seen on the example of the analysed building, currently under construction in Warsaw city centre – an almost 200 m tall building, irregular and asymmetrical shape with numerous facade recesses and geometry based on right angles. The tower is formed of slender blocks of various heights and widths (Figure. 1). Functionally and spatially separated there are: a 10-level plinth part and a 36-level high dominant. The shape of the ground floor projection is similar to an elongated rectangle measuring 85 x 35 m in the plinth part and 65 x 30 m in the tower part.

The tower is located in an intensive development zone. On the northeast side (Figure. 2) there is a complex of multi-level buildings. On the east side, low and medium-high buildings prevail. On the south side there are some low-rise buildings. On the north side, a tall building complex is being developed, with another being built from the west.

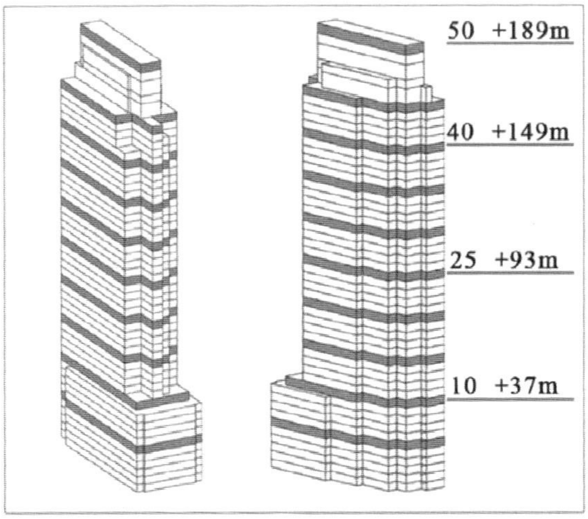

Figure 1. Model of the analysed building. Highlighted: floors selected for the analyses and presented in the article.

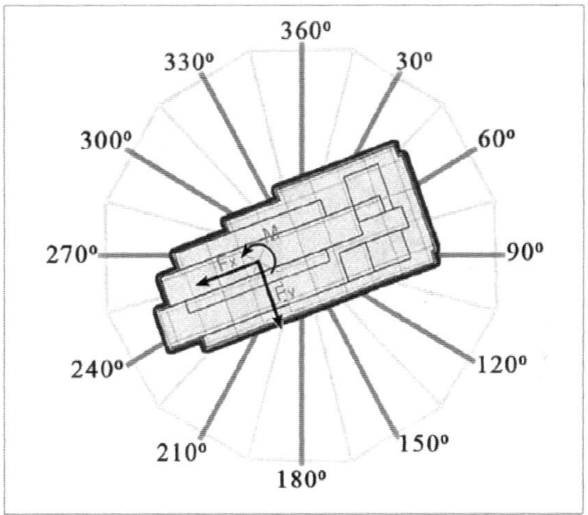

Figure 2. Analysed building – the coordinates system, geographical and wind directions.

3. Description of the study

The aim of the undertaken aerodynamic studies was to compare the results obtained while using two different methods: the ‘European standard method’, which consisted in carrying out calculations in accordance with applicable regulations and standards, and the empirical (experimental) method included wind tunnel experiments.

The study compared the results obtained for 10 selected floor levels (Figure. 1) and all façades. The results presented in this article were limited to 4 selected floor levels (some are given only for floor 25) and the southern façade.

IOP Conf. Series: Materials Science and Engineering **809** (2020) 012006 doi:10.1088/1757-899X/809/1/012006

The analyses were conducted for 12 wind directions (Figure. 2) (because in the Eurocode, the variation of the directional coefficient c_{dir} is included for 12 directions). This paper presents the comparisons in terms of: pressure distribution on façades, envelopes and magnitude of peak pressures, and the magnitude of the net forces F_x and F_y transferred to the structure at the selected floor levels.

4. Methods

4.1 PN-EN 1991-1-4:2008 1 Eurocode 1.

The 'European standard method' included the analysis of building load conditions according to PN-EN 1991-1-4:2008 Eurocode 1 (described in [*9*]) and in accordance with Eurocode guidelines did not take into account the detailed configuration of the surrounding buildings, only a very generally described terrain class. According to the Polish annex to Eurocode, 1 wind load zone and IV category of terrain were assumed. The atmospheric boundary layer characteristic was mapped in all methods by adopting a standard mean wind velocity and turbulence intensity profiles.

4.2 Wind tunnel testing.

The empirical method consist in experimental research in a closed-loop wind tunnel carried out at the Institute of Aeronautics and Applied Mechanics at the Warsaw University of Technology (described in [*9*]). Passive methods were used to map the atmospheric boundary layer characteristic in the tunnel. The examined building, together with neighbouring buildings within a 500 m radius, was modelled on a scale of 1:350 (Figure. 3). The experiments were carried out with the use of rigid building models that allow measuring wind pressure on the walls using pressure sensors, as well as measuring net forces and aerodynamic moments using an aerodynamic balance. The signals from the measurements have been processed numerically.

Figure 3. Model of the analysed building and its surroundings in the wind tunnel.

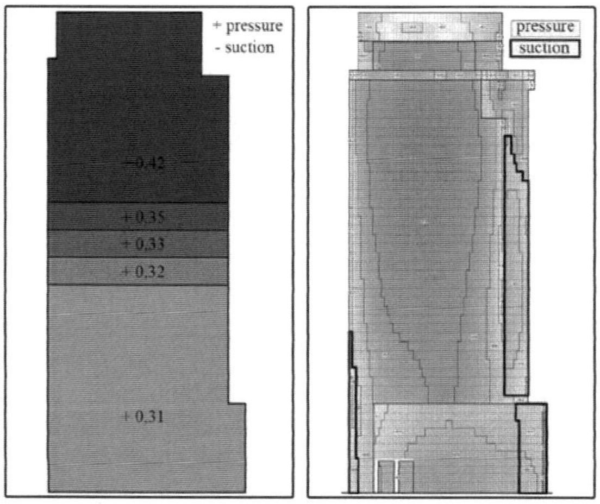

Figure 4. Maps of the wind pressure distributions on the southern façade, obtained according to: Eurocode-based procedure and wind tunnel testing. 160° wind direction.

5. Findings

5.1 Peak pressure values.

The first analysis concerned the distribution of pressure on the facades (Figure. 4). To limit the impact of the environment, the analysis has been restricted to comparing pressure distributions on the windward southern façade (160° wind direction). The results obtained in the Eurocode-based analysis are significantly inaccurate. The differences in pressure distribution resulting from the irregular shape of the structure are not visible. Moreover, the occurring changes in pressure values are largely simplified in relation to reality. On the other hand, in wind tunnel experiments it is possible to obtain highly precise results that could form the basis of a detailed optimization of a building with a complicated form

Afterwards, peak pressure values were analysed. Comparing peak pressure values for long façades (north and south) (Figure. 5, Table 1 and 3), one can see that the values obtained from the Eurocode-based calculations can be twice as high as those measured in the wind tunnel. For the shorter eastern and western facades and around corners and bends the results are similar (Figure. 6, Table 2 and 4). Strict Eurocode-based guidelines regarding the edges and bends of the building, with respect to short or broken-up facades, have been experimentally confirmed.

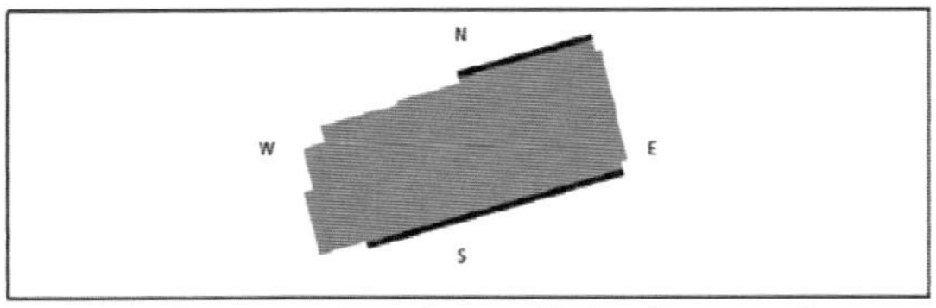

Figure 5. Façades for which the results are presented in Tables 1 and 3.

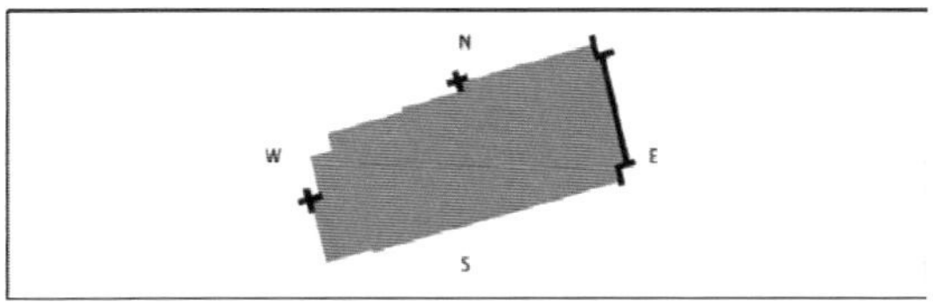

Figure 6. Façades and corners for which the results are presented in Tables 2 and 4

Table 1. Peak pressure values for the northern façade (EC – Eurocode, TT – Tunnel Test)

Floor	Pressure [kPa]		Suction [kPa]	
	EC	TT	EC	TT
10th	0,83	0,37 – 0,53	0,55 – 1,02	0,41 – 0,60
25th	0,85	0,40 – 0,62	0,57 – 1,05	0,37 – 0,73
40th	1,12	0,64 – 0,75	0,74 – 1,38	0,38 – 0,70
50th	1,12	0,61 – 1,00	0,74 – 1,38	0,27 – 0,44

Table 2. Peak pressure values for the western façade (EC – Eurocode, TT – Tunnel Test)

Floor	Pressure [kPa]		Suction [kPa]	
	EC	TT	EC	TT
10th	0,50	0,27 0,50	0,83 – 1,02	0,73 – 1,03
25th	0,52	0,35 – 0,55	0,83 – 1,05	0,80 – 1,00
40th	0,68	0,39 – 0,54	1,09 – 1,38	0,39 – 0,54
50th	0,68	0,40 – 0,50	1,38	0,35 – 0,66

Table 3. Peak pressure values for the southern façade (EC – Eurocode, TT – Tunnel Test)

Floor	Pressure [kPa]		Suction [kPa]	
	EC	TT	EC	TT
10th	0,65	0,20 – 0,39	0,57 - 1,02	0,55 – 0,69
25th	0,67	0,29 – 0,44	0,59 – 1,05	0,54 – 0,70
40th	0,88	0,36 – 0,48	0,78 - 1,38	0,60 - 0,65
50th	0,88	0,35 – 0,66	0,78 - 1,38	0,36 – 0,70

Table 4. Peak pressure values for the chosen corners on the 25th floor (EC – Eurocode, TT – Tunnel Test)

Corner	Pressure [kPa]	
	EC	TT
W	1,05	1,09
N	1,29	1,27
N-E	0,63	0,58
S-E	0,63	0,73

One can see that the effect of suction is particularly important here. Much higher values are observed on virtually all corners, rather than on flat sections. For the western facade, the Eurocode-based method does not reflect the differences in the amount of wind suction. The differences between the Eurocode-based values and values obtained from the wind tunnel tests are the largest for the southern façade and flat surfaces of the northern façade. When it comes to wind pressure according to Eurocode, the values remain identical over the entire length of the façade, while the wind tunnel analysis results show some clear differences.

5.2 Net Forces.

The study analysed the F_x and F_y components of net forces for selected floors combined for a fixed point in the building's coordinate system (Figure. 2). Significant differences in the results were observed: usually the results obtained from the Eurocode-based calculations were 2-3 times greater than those measured in the tunnel (e.g. F_y component for 340°direction – Table 5), and in extreme cases the differences can be 7-10 times greater (70° direction – F_x component for the 50th floor and F_y component for the 10th floor – Table 6), and the component forces can even have an opposite direction (F_x component, direction 340° - Table 5).

Table 5. The components F_x and F_y of net forces for selected floors – 340° wind direction (EC – Eurocode, TT – Tunnel Test)

Floor	Fx [kN]		Fy [kN]	
	EC	TT	EC	TT
10th	-98,06	13,80	330,25	175,80
25th	-72,88	26,50	245,46	75,90
40th	-66,43	15,30	223,72	74,80
50th	-70,79	-4,90	211,93	66,10

Table 6. The components F_x and F_y of net forces for selected floors – 70° wind direction (EC – Eurocode, TT – Tunnel Test)

Floor	Fx [kN]		Fy [kN]	
	EC	TT	EC	TT
10th	75,25	26,30	-106,58	-10,20
25th	55,93	16,00	-79,21	-11,30
40th	50,98	15,10	-72,20	-25,50
50th	43,62	6,80	-68,17	-28,50

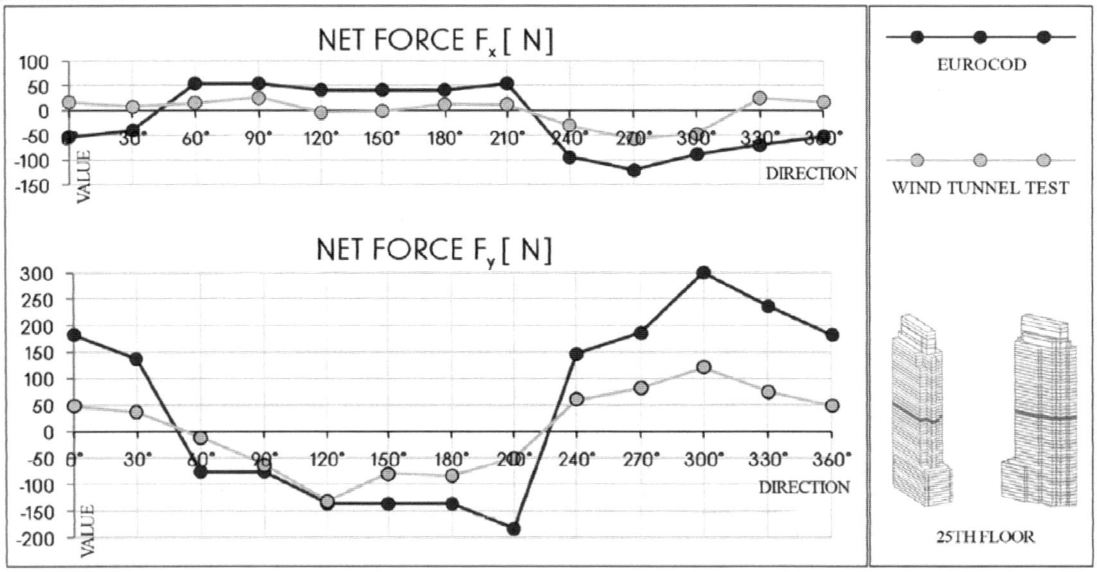

Figure 7. The components F_x and F_y of net forces for the 25th floor, depending on the direction of wind

By analysing the chart of the components of net forces, e.g. for the 25th floor (Figure. 7), the impact of surrounding buildings on the magnitude of wind forces can be clearly observed. The most varied are the net forces calculated for winds blowing from the north (300°, 340°, 0°) and west (210°, 240°, 270°), which results, among others from close proximity to high-rise buildings. On the other hand, smaller differences in the net forces for the south-eastern wind directions (120°, 150°, 180°) result from the lack of objects of significant height in this area in the immediate vicinity. However, the position of the building –

perpendicular to the direction of wind – results in a greater variation in the results obtained. For east winds (60°, 90°), the results are even more convergent. There are no tall constructions on this side, the wind blows parallel to the longer side of the analysed building and its form can be treated as more streamlined.

6. Summary

The accuracy of the results obtained in wind tunnel analysis testing allows for a fairly precise determination of pressure distributions and net forces transmitted to the structure. During the analysis of pressures on building façades, the possibility to optimize based on the results of experimental research was determined. Particularly large differences were noticeable in the analysis of net forces acting on the building. The net forces values measured experimentally were much smaller than those calculated in accordance with Eurocode. Not only the impact of the complex geometry of the building, but also the influence of neighbouring high-rise buildings and the resulting aerodynamic shadows are noticeable.

7. Conclusions

Due to significant wind load effects on tall buildings, optimization of their form requires to identify those loads as accurately as possible. Determining methods for conducting wind analysis becomes a key issue, as the results obtained can differ significantly. The analyses confirmed the hypothesis that results obtained with different methods are divergent and the choice of method has a significant impact on the adopted technical solutions.

The analysed methods differ in the accuracy of the results obtained. Wind-tunnel testing allow more accurate analysis of the wind load effect on objects located in a complex environment, and especially those with unusual geometry. The European Standard procedures have been developed to analyse simple, basic geometries, not taking into account the impact of the surrounding environment, and the results obtained often lead to the adoption of excessive safety factors. A simplified representation of the wind load nature in relation to their value and distribution also results in under clarification and overestimation of the values calculated.

Although the European Standard can be applied to buildings up to a height of 200 m, the procedures described are only sufficient for calculating wind loads of simple, regularly shaped buildings. For buildings with complex, irregular geometry, the Standard does not specify any more accurate procedures. Because the Eurocode-based methods do not adequately analyse many aspects of structural design, the safety factors assumed are too high. On the other hand, the results obtained from experimental tests, as more reliable, can provide a reference point for accepting lower wind load values. It should be noted, however, that structural design process based on the Eurocode Standard is a safe procedure and accepting lower load than those described in the Eurocode should be each time justified by detailed analytical analyses.

Acknowledgments

This research was carried out as a part of grant at the Warsaw University of Technology, Faculty of Architecture.

8. References

[1] Boda, D. and Banda, L. (2011). *Wind effects on typical tall structures*. Saarbrücken: Lambert Academic Publishing.

[2] Irwin, P., Denoon, R. and Scott, D. (2013). *Wind Tunnel Testing of High-Rise Buildings: An output of the CTBUH Wind Engineering Working Group*. Chicago: Council on Tall Buildings and Urban Habitat, Routledge / Taylor and Francis Group.

[3] Kahn, F. R. (1982). The rise and fall of structural logic in architecture. *Chicago Architectural Journal*

[4] Billington, D. (1983). *The tower and the bridge: The new art of structural engineering*, Princeton University Press, Princeton.

[5] Flaga, A. (2008). *Inżynieria wiatrowa. Podstway i zastosowania.* Warszawa: Arkady.

IOP Conf. Series: Materials Science and Engineering **809** (2020) 012006 doi:10.1088/1757-899X/809/1/012006

[6] PN-EN 1991-1-4:2008 Eurokod 1: oddziaływania na konstrukcje. Część 1-4, Oddziaływania ogólne: oddziaływania wiatru.
[7] Żurański, J. A. and Gaczek, M. (2011). *Oddziaływania klimatyczne na konstrukcje budowlane według Eurokodu 1. Komentarz z przykładami obliczeń.* Warszawa: Instytut Techniki Budowlanej.
[8] Lawson, T. (2001). *Building aerodynamics.* Londyn: Imperial College Press.
[9] Pietrzak, J., Rutkowski M. and Wrona M. (2017). *Analizy normowe oraz badania modelowe oddziaływania wiatru na budynek wysoki o nieregularnej formie na przykładzie projektowanego wieżowca w Warszawie.* Unpublished manuscript, Politechnika Warszawska, Wydział Architektury, Warszawa.

Experimental Investigation for Moment Redistribution in Continuous RC Beams Top Strengthened with CFRP and Steel Plates

Amr Ibrahim[1,a], Shady Salem[1,b], Ayman Khalil[2] and Mahmoud El-Kateb[2]

[1]The British University in Egypt/Civil Engineering Department, Cairo, Egypt
[2]Ain Shams University/Civil Engineering Department, Cairo, Egypt

[a]amr.ibrahim@bue.edu.eg, [b]shady.salem@bue.edu.eg

Abstract. An experimental study was conducted on eight T-section continuous reinforced concrete beams top strengthened with CFRP and steel plates. The beams had a constant cross section and flexural steel reinforcement which was designed to prevent brittle failure due to crushing of concrete before steel yielding. The plates were installed on both sides of the slab above the beam web as near surface mounted or externally bonded to investigate the effect of different plating techniques on the moment redistribution in reinforced concrete beams. The number of plates in each specimen was changed. Steel stirrups were used and designed so that no premature shear failure occurs. The experimental results show that the steel plating can cause up to 40 % redistribution of moments from the sagging to the hogging regions compared to the control specimen as well as a significant increase in the loading capacity of the beams. The results also revealed that the application of CFRP plates can lead to substantial amounts of moment redistribution when they are installed as near surface mounted plates, but without enhancing the load carrying capacity.

1. Introduction

Strengthening of Reinforced Concrete (RC) elements using different materials and techniques is widely used to avoid deterioration of existing structures or to increase the capacity of an RC element along a certain cross-section. Regarding continuous RC members, strengthening at a given section can cause a significant amount of moment redistribution (MR). The concept of MR in RC elements is identified in many codes such as ACI 318-14 [1] and CSA/A23.3-14 [2]. When using conventional steel as reinforcement, the yielding of steel results in most of the MR; also several studies [3,4] reported that after the formation of cracks and before yielding of steel at the critical sections, moment redistribution is likely to occur due to the difference in flexural stiffness along the beam.

MR in continuous members allows more flexibility in structural design, which is usually carried out by reducing the hogging moments over supports [5]. Correspondingly, moments at the sagging areas increase to achieve equilibrium. However, in some cases, MR is required to be achieved by increasing the hogging moments at the expense of the sagging moments.

For example, the designers may consider the MR for strengthening only the upper side of continuous structural elements, beams or slabs, in case of having inaccessibility for strengthening the bottom surface of those elements. This strengthening technique is based on increasing the moment over supports and reducing the moment at mid-span due to the modified stiffness of the strengthening system. Strengthening of the hogging moment areas can be achieved either using steel or carbon fiber reinforced polymers (CFRP). Strengthening with steel plates/rebars are characterized with their high ductility which allows for significant values of MR to occur. Steel strengthening also has sufficient bond strength with concrete which ensures that MR can be achieved to the maximum possible limit without debonding [6,7].

On the other side, Carbon Fiber Reinforced Polymers (CFRP) is considered as a good alternative to conventional steel due to the non-corrodible nature and high strength to weight ratio. However, CFRP does not undergo yielding process rather they tend to behave in a linear-elastic manner until failure [8], this behavior affects the ability of RC members strengthened with CFRP to achieve MR between the critical sections. Generally, when strengthening a structural element using FRP material, the failure mode tends to be more brittle which reduces MR as the ductility is a key factor to redistribute moments in a continuous RC member. Moreover, several studies showed that CFRP plates often debond earlier than steel plates [6,7,9].

Furthermore, retrofitting of beams and slabs can be achieved by using externally bonded (EB) plates or near-surface mounted (NSM) CFRP or steel plates. Research has shown that EB plates tend to debond at relatively low strains [10,11,12]. As such, the current guidelines neglect the moment redistribution in plated RC flexural members. On the contrary, recent research on retrofitting using FRP and steel NSM plates has shown that NSM plates tend to debond at high strains which can allow for significant amounts of MR to occur [13,14].

As such, the presented study summarizes the experimental investigation for a novel MR approach for RC continuous beams strengthened at the hogging zones only using conventional steel as well as CFRP plates. The method of installment for those plates will vary from EB plates at the top surface of the beams to NSM plates buried in the upper concrete cover of the tested beams.

2. Experimental Program

Eight two-span continuous beams with a span length of two meters were tested using two concentrated loads loaded at their mid-spans as shown in Figure 1. The loads were applied in a monolithic static way until failure. Figure 1 also demonstrates the typical reinforcement details of the tested beams. The beams were designed as T-sections, as shown in Figure 2, to increase the flexural capacity of the section at mid-span between the supports and to avoid early failure of the beams at mid-span sections. Figure 2 also discloses the positions of the used plates in each strengthening technique (1500 mm at the location of the intermediate support). The used plates were implemented at the flanges of the beam within the effective width of the slab rather than in the web of the beam to avoid the presence of an intermediate column if any.

Table 1 summarizes the test matrix and demonstrates the tested parameters. The steel plates were 50 mm in width, 6 mm in thickness, while the CFRP laminates were 50 mm in width, 1.2 mm in thickness. The steel plates and the CFRP laminates were coated by an epoxy adhesive material. The tested beams were named after the type (Steel or CFRP), method of installation number (EB or NSM), and of the number of the used plates. For example, beam CN1 is strengthened with one NSM CFRP plate. For the sake of clarity, the control beam was named “B1”. Figure 3 shows that all the beams were instrumented with six Linear Variable deferential transducers (LVDT) and nine strain gauges (three measuring the concrete compression strain (CS1 to CS3), five gauges measuring the main lower and upper reinforcement strains (SS1 to SS5), and one strain gauge measuring the strengthening plate strain (SS6). Figure 4 summarizes the typical locations of the used instrumentation.

3. Material Properies

The characteristic compressive strength of concrete was 28.7 MPa, with a coefficient of variation (COV of about 9.8%) based on 18 cubes tested according to the British Standards [15]. Moreover, the yield and ultimate tensile strength of the main reinforcement of the steel beams were 360 and 520 MPa respectively, while the yield and ultimate tensile strength of the strengthening steel plates were 425 and 566 MPa with percentage of elongation of 21.9%. The ultimate tensile strength of the CFRP plates used (Sika Carbodur S-512) was 3100 MPa based on the manufacturer’s data, while the modulus of elasticity was 165 GPa with mean percentage of elongation of 1.8%

4. Failure Modes

Generally, cracks first appeared at the load application points and then, for higher loads, at the central support. The control beam underwent an ordinary flexural failure at one of the two load application sections with the yielding of the steel reinforcement. For the control beam, initial cracks were observed at a load level of 160 kN at the positive moment zone, then narrow cracks appeared at the negative moment zone at 300 kN. The failure commenced at the sagging regions at (P = 435 kN) through concrete cracking as depicted in Figure 4. The behaviors of the strengthened beams SN1, SE1, CN1, and CN2 were close to the control specimen in terms of the crack initiation, pattern and failure mode while having different failure loads as shown in Figure 5.

On the contrary, beams SN2 and SE2 failed at one of the sections near the load application points by crushing the concrete at the bottom side and at the middle support section as shown in Figure 6. Initial hair cracks appeared at the sagging zone at P = 80 kN, while cracks began to be visible at the central support at P = 220 kN. The cracks began to widen and spread almost simultaneously until failure.

Regarding CE1, the cracks first appeared under the load application point at P = 80 kN, however, the cracks were noticed at the central support at P = 160 kN. At P = 360 kN the CFRP start to debond as shown in Figure 7. After debonding, the beam continued to bear more loads but with the rapid increase of the vertical displacement under the load application point. Then the applied load remained nearly cc stopped at Pu = 404.71 mm.

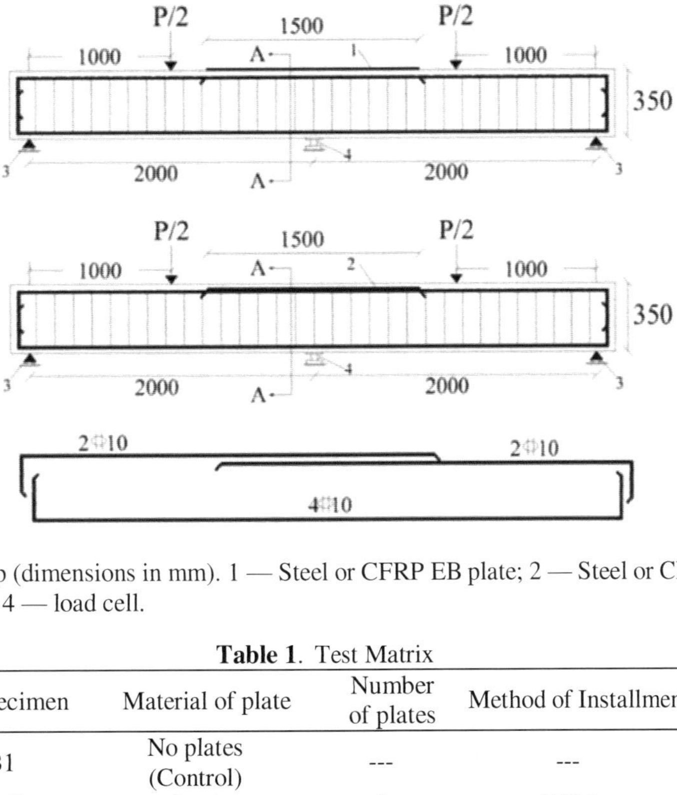

Figure 1. Test setup (dimensions in mm). 1 — Steel or CFRP EB plate; 2 — Steel or CFRP NSM plate; 3 — roller support, 4 — load cell.

Table 1. Test Matrix

Specimen	Material of plate	Number of plates	Method of Installment
B1	No plates (Control)	---	---
SN1	Steel	1	NSM
SE1	Steel	1	EB
SN2	Steel	2	NSM
SE2	Steel	2	EB
CN1	CFRP	1	NSM
CE1	CFRP	1	EB
CN2	CFRP	2	NSM

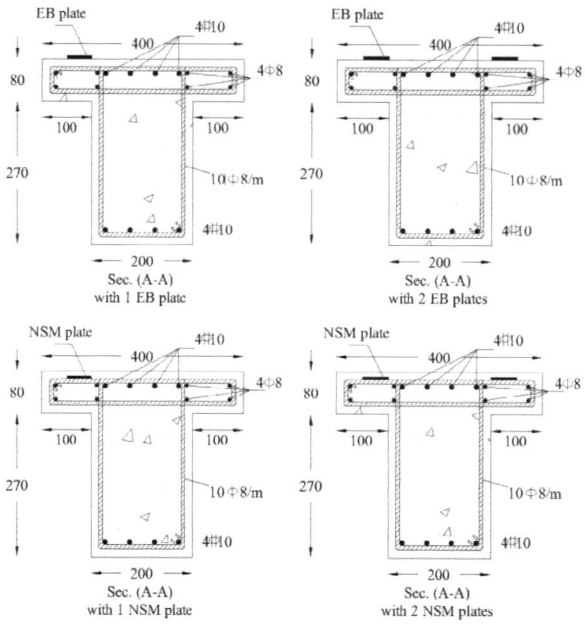

Figure 2. Cross-sections (dimensions in mm)

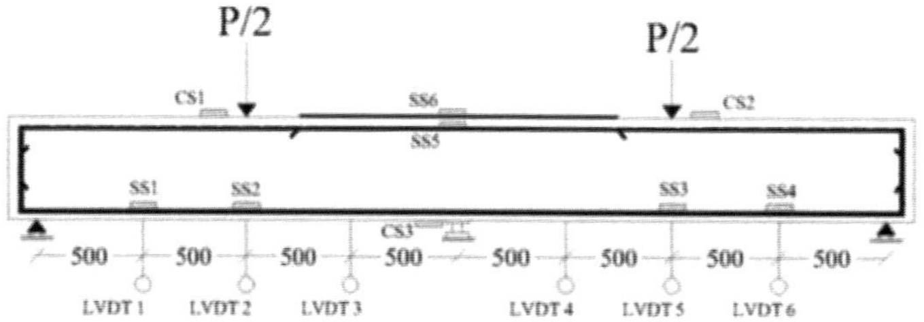

Figure 3. Typical instrumentation of the tested beams

5. Test Results

Table 2 summarizes the test results, where P_u is the maximum load applied on the beam, R_m is the vertical reaction of the central support at load P_u, R_e is the end support reaction, while, M_{sag} and M_{hog} are the experimental flexural moments at the positions of mid-span and at the central support respectively. It is worth mentioning, that the own weight of the beams was neglected in the calculations. The percentages of increase in the load capacity of the beam, reduction of M_{sag} and increase in M_{hog} were calculated relative to the control specimen.

The load capacity and the percentage of MR for SN1 and SE1were similar, while SN2 had higher capacity and MR than SE2. For example, SN2 had 7 and 11% more in terms of capacity and reduction in hogging moment respectively than SE2. Furthermore, SN2 showed higher load capacity, reduction in M_{sag}, increase in M_{hog} than SN1 by about 17, 8, and 31% respectively. Moreover, SE2 showed higher load capacity, reduction in M_{sag}, increase in M_{hog} than SE1 by about 1.5, 3.5, and 19% respectively. As such, it can be observed that the application of NSM plates was more effective than EB plates when using steel plates as was shown in the comparison between SN2 and SE2.

For CE1, debonding of the CFRP plate from the concrete surface occurred at Pu = 360 kN, however, the beam continued to bear extra load until the failure load of 404.71 kN, which resulted in a negligible MR at the maximum load, however, some MR was observed during loading of the beam before debonding. Similarly, the ultimate capacity of CN1 did not increase, however, MR was observed at the sagging and hogging moment locations. CN2 showed MR of 15% more than CN1. As such, The application of CFRP NSM plates proved to be more effective than EB plates and the usage of two plates gave higher percentages of moment redistribution. However, the beams strengthened with CFRP plates gave smaller values of increased load capacity and MR than the beams strengthened with steel plates.

Table 2. Test Results

Beam	P_u (kN)	R_m (kN)	R_e (kN)	M_{sag} (kN.m)	M_{hog} (kN.m)	[%] of load capacity increase	[%] of reduction, M_{sag}	[%] of increase, M_{hog}	Average [%] of MR
B1	467.19	360.40	53.40	53.40	-126.81	N.A.	N.A.	N.A.	N.A.
SN1	480.86	391.53	44.67	44.67	-151.10	2.93	16.35	19.16	17.75
SE1	475.24	387.25	43.88	43.88	-149.75	1.67	17.83	18.09	17.96
SN2	561.37	478.95	41.21	41.21	-198.27	20.16	22.82	56.35	39.59
SE2	525.72	441.12	42.30	42.30	-178.26	12.53	20.78	40.58	30.68
CN1	462.27	369.84	46.22	46.22	-138.71	-1.05	13.45	9.38	11.42
CE1	404.71	315.72	44.50	44.50	-113.37	-13.37	16.67	-10.60	3.03
CN2	468.55	380.41	44.07	44.07	-146.14	0.29	17.46	15.24	16.35

Figure 4. Failure mode for beam (B1)

Figure 5. Typical failure mode for beams SN1, SE1, CN1 and CN2

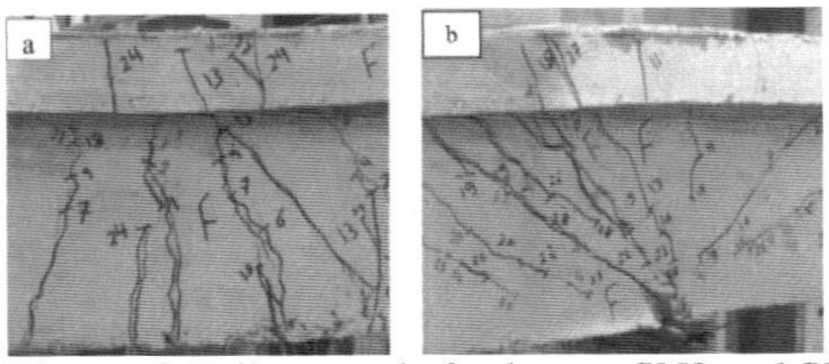

Figure 6. Failure mode for beams SN2 and SE2 a. at mid-span; b. at the central support

Figure . 7. Debonding of the CFRP plate in beam CE1

6. Analysis of Results

6.1. Strains.

Figure 8 demonstrates the tensile strain of the lower reinforcement, showing a similar trend for all the beams except SN2 and SE2 (i.e. with small yielding plateau). This different behavior supports the fact that these beams had smaller ductility controlled by the concrete crushing. It was also noticed that the application of CFRP plates did not affect the behavior of the load-strain relationship in the linear elastic stage. On the other side, the concrete compressive strains measured at the top side at the mid span sections or the lower side at the middle support location did not exceed 0.003 as shown in Figure 9.

The control beam yielded at the lowest load level which is mainly because of that the positive flexural stresses at mid-span caused by sagging moments. Generally, the yielding loads increased with the decrease in the sagging moments in the beams which were clear in the behavior of the rest of the beams, except for CN2 which had a sagging moment less than CN1, SE2, and SN2 but still had a lower yielding load than those beams.

6.2 Deflection.

Figure 10 shows the load-displacement curve at the point of load application. From the figure, both beams B1 and CE1 showed large ultimate deformation, 27.15 mm and 29.75 mm for B1 and CE1

IOP Conf. Series: Materials Science and Engineering **809** (2020) 012007 doi:10.1088/1757-899X/809/1/012007

respectively. The similar behavior of beam CE1, behaved to B1, is attributed to the debonding of the CFRP plate which lead to a non-strengthened beam behavior as beam B1.

It is also noted that the application of CFRP strengthening did not affect the flexural stiffness of the beams considerably in the linear elastic range as it is shown that all the beams behaved in an elastic behavior at the beginning until reached a value between 300 and 360 KN, after that, the behavior of the beams was varying depending on the strengthening technique.

All the strengthened beams, except for CE1, had lower values of the ultimate displacement compared to the control specimen. This is mainly due to the redistribution of moments from sagging to hogging zones, which means that the reduction of sagging moments resulted in less displacement values

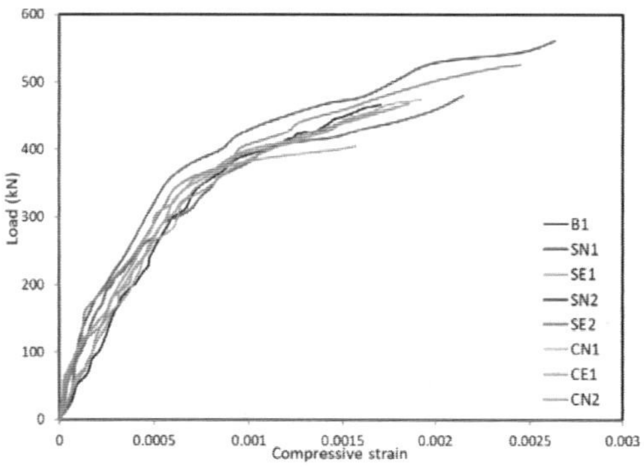

Figure 8. Applied load versus maximum measured tensile strains of lower reinforcement

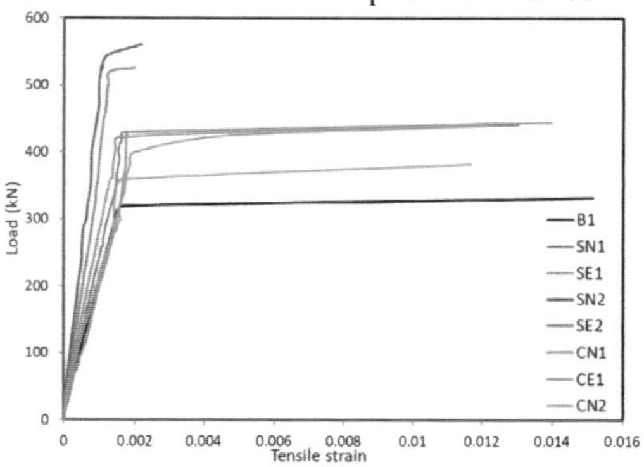

Figure 9. Applied load versus compressive strain of the bottom side of concrete at the central support

Table 3. Deflection Ductility Indices

Beam	Δ_u (mm)	Δ_y (mm)	Δ_{cr} (mm)	$\mu_{\Delta u\text{-}cr}$	$\mu_{\Delta u\text{-}y}$
B1	27.15	4.55	4.74	5.73	5.97
SN1	19.36	10.92	4.68	4.14	1.77
SE1	17.13	8.39	3.75	4.57	2.04
SN2	19.61	15.42	5.45	3.60	1.27
SE2	18.63	15.58	4.11	4.53	1.20
CN1	15.93	10.06	4.33	3.68	1.58
CE1	29.75	6.65	4.22	7.05	4.47
CN2	14.44	8.58	4.14	3.49	1.68

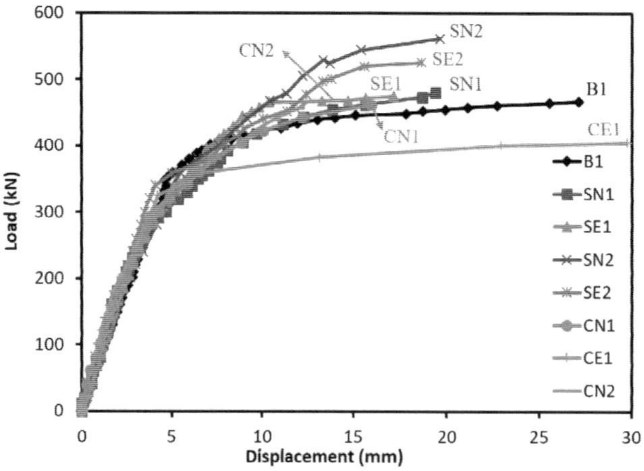

Figure 10. Applied load versus maximum displacement at points of load application for each beam

6.3 Ductility Indices.

Table 3 shows the deflection ductility indices including the ratio of the ultimate to yield deflection, which is known as the ductility factor and is often used as a simple measure of energy absorption [16]. The deflection ductility indices $\mu_{\Delta u\text{-}cr} = \Delta_u / \Delta_{cr}$ and $\mu_{\Delta u\text{-}y} = \Delta_u / \Delta_y$ which are the ratios of the ultimate deformation (Δ_u) to the deformation at the onset of cracking (Δ_{cr}) and the deformation at the point of yielding of the lower steel reinforcement in the sagging zones (Δ_y). All the strengthened beams had smaller values of $\mu_{\Delta u\text{-}y}$ than the control specimen, regardless the amount of moment redistribution. Thus, it can be concluded that the moment redistribution is related not only to the capacity of inelastic deformation but also to a proper design of the ultimate capacity of critical sections.

Beam CE1 recorded the highest value of $\mu_{\Delta u\text{-}y}$ since it acted as a non-strengthened beam for the late stages of the test after debonding of the CFRP plate. This is so close to the case with $\mu_{\Delta u\text{-}cr}$ in which the values determined for the control specimen and CE1 are the highest among all the experimented specimens.

6.4 Load vs percentage of moment redistribution.

As shown in figures 11 and 12, the initial values of moment redistribution were not negligible because the flexural stiffness of the beam in the hogging region (strengthened zone) was higher than in the sagging region. For beam SN2, the percentage of reduction in the sagging moment decreased from the initial values they started with a high rate until the cracks appeared near the central support, when a slight rise occurred and then the reduction decreased with a very low rate until the end of the test to show eventually a final value almost equal to the value occurred at the appearance of the cracks near the intermediate support. For SE2, the values remained almost constant during the test and achieved lower rates than that achieved by SN2 at all load levels.

The behavior of beam CN1 was close to that of SN2 in which the percentages of MR decreased rapidly at first until the moment when the cracks appeared near the central support, and then the values decreased with a smaller rate until the lower steel reinforcement at the sagging regions yielded at p = 422.75 kN before the values started to rise given a final value close to what was achieved when the first cracks appeared near the central support.

For CN2, the behavior was similar to that of CN1, the beam had the highest initial MR percentages. These values decreased rapidly until the cracks appeared at the central support when a slight rise took place, then the values continued in decreasing at a high rate until a load level of about 400 kN. That is when the lower steel bars at the sagging zones reached yield and the values began to increase until failure.

Beam CE1 witnessed an initial decrease of MR percentage until the load level of about 250 kN when a significant rise in the values occurred until the load level of 360 kN which is when the CFRP plate was debonded from the concrete surface. After that, the values decreased abruptly until failure at Pu = 404.71 kN.

The behavior of beams SN1 and SE1 was different from that of the previous ones. The MR percentages started with negative values and increased during loading until the load level of about 340 KN. After that, the values remained constant until the reinforcement in the sagging zones reached yield at load levels approximately equal to 430 kN before the values rose again reaching significant positive values at the end.

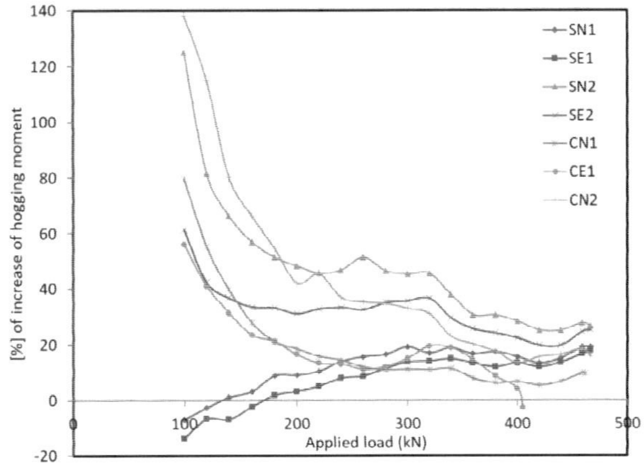

Figure 11 Applied load versus the percentage of reduction in sagging moments for all beams

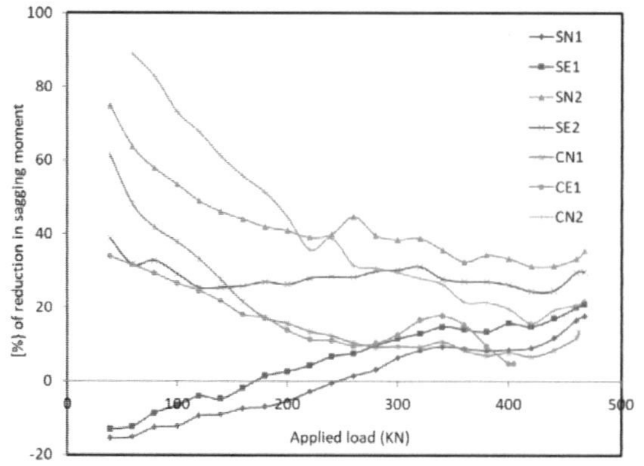

Figure. 12 Applied load versus the percentage of increase in hogging moments for all beams

7. Conclusions

The NSM steel plating technique at the hogging zones caused up to 40 % redistribution of moments from the sagging to the hogging regions compared to the control specimen. Moreover, the NSM plates were found more effective than the EB ones when two plates were used as SN2 had 7 and 11% more in terms of load carrying capacity and decrease in hogging moment respectively than SE2. However, the difference between the aforementioned techniques was negligible when only one plate was used.

The application of two steel plates was more efficient than one steel plate as SN2 showed higher load capacity, reduction in sagging moment, increase in hogging moment than SN1 by about 17, 8, and 31% respectively. Moreover, SE2 showed higher load capacity, reduction in M_{sag}, increase in M_{hog} than SE1 by about 1.5, 3.5, and 19% respectively.

Furthermore, the application of CFRP plates can lead to substantial amounts of moment redistribution when they are installed as NSM plates, but without enhancing the loading capacity.

Significant redistribution of moments occurred in the tested beams prior to the plastic limit being reached in either the reinforcement or the concrete. This elastic redistribution took place as a result of the variation in flexural rigidity throughout the span because of the presence of the plates and due to cracking at the critical sections. The application of CFRP strengthening techniques investigated in this study did not affect the flexural stiffness of the beams considerably in the linear elastic range.

Acknowledgment
The first author would like to thank the British University in Egypt for their financial support.

8. References

[1] ACI 318, Building code requirements for reinforced concrete and commentary, ACI 318-14, American Concrete Institute, Detroit, MI, 2014.
[2] CSA/A23.3, Canadian Standard Association, Code for the design of concrete structures for buildings, CSA/A23.3-14, Toronto, Ontario, Canada, 2014.
[3] H. Scholz, "Contribution to redistribution of moments in continuous reinforced concrete beams," ACI Structural Journal vol. 90 (2): pp. 150-155, 1993.
[4] R.N.F. Carmo and S.M.R. Lopes, "Available plastic rotation in continuous high-strength concrete beams," Canadian Journal of Civil Engineering, vol. 35 (10), pp. 1152-1162, October 2008.
[5] I.F. Kara and A.F. Ashour, "Moment redistribution in Continuous FRP reinforced concrete beams," Construction and Building materials, vol. 49. pp. 939-948, December 2013.
[6] D.J. Oehlers, G. Ju, I.S.T. Liu and R. Seracino "Moment redistribution in continuous plated RC flexural members. Part 1: neutral axis depth approach and tests," Engineering Structures, Vol 26, pp. 2197-2207, 2004.
[7] D.J. Oehlers, G. Ju, I.S.T. Liu and R. Seracino "Moment redistribution in continuous plated RC flexural members. Part 2: flexural rigidity approach", Engineering Structures, Vol 26, pp. 2209-2218, 2004.
[8] S.M.H. Rahman, K. Mahmoud, and E. El-Salakawy, "Moment redistribution in glass fiber reinforced polymer-reinforced concrete continuous beams subjected to unsymmetrical loading," Engineering Structures, vol. 150. pp. 562-572, 2017.

IOP Conf. Series: Materials Science and Engineering **809** (2020) 012007 doi:10.1088/1757-899X/809/1/012007

[9] D.J. Oehlers, S.M. Park and M.S. Mohamed Ali, "A Structural Engineering Approach to Adhesive Bonding Longitudinal Plates to RC Beams and Slabs," Composites Part A: Applied Science and Manufacturing, vol. 34 (9), pp. 887–897, 2003.
[10] I.S.T. Liu, D. J. Oehlers and R.Seracino, "Moment Redistribution in FRP and Steel-Plated Reinforced Concrete Beams," Journal of Composites for Construction", vol. 10(2), 2006.
[11] A.F. Ashour, S. A. El-Refaie, and S.W. Garrity, "Flexural strengthening of RC continuous beams using CFRP laminates," Cement and Concrete Composites, vol. 26, pp. 765–775, 2004.
[12] S.A. El-Refaie, A.F. Ashour, and S.W. Garrity, "Sagging and hogging strengthening of continuous reinforced concrete beams using carbon fiber-reinforced polymer sheets," ACI Structural Journal, vol. 100(4), pp. 446–453, 2003.
[13] L. De Lorenzis, A.Nanni, and C. Modena, "Bond between near-surface mounted fiber-reinforced polymer rods and concrete in structural strengthening," ACI Structural Journal, vol. 99 (2), pp.123–132, 2002.
[14] T. Hassan, and S. Rizkalla, "Investigation of bond in concrete structures strengthened with near-surface mounted carbon fiber reinforced polymer strips," Journal of Composites for Construction, vol. 7 (3), pp.248–257, August 2003.
[15] BS EN12390-3, "Testing hardened concrete part 3: Compressive strength of test specimens," British standards, 2009.
[16] R.N.F. Carmo and S.M.R. Lopes, "Ductility and linear analysis with moment redistribution in reinforced high-strength concrete beams", Canadian Journal of Civil Engineering, vol. 32 (1), pp. 194–203, 2005.

Improving the Effectiveness of Stack Ventilation by Supplying an Outdoor Air Stream

Katarzyna Gładyszewska-Fiedoruk[1,a] and Manuel Ruiz de Adana[2,b]

[1]Bialystok University of Technology, Department of HVAC Engineering, Bialystok, Poland

[2]University of Cordoba, Department of Applied Thermodynamic, Cordoba, Spain

[a]k.gladyszewska@pb.edu.pl, [b]manuel.ruiz@uco.es

Abstract. Stack ventilation is the most common ventilation in public utility buildings and residential homes. Its task is to remove contaminated air from the room and thus maintain good indoor air quality. The aim of the study was to determine the effectiveness of the natural ventilation of one classroom and determine the impact of the additional air supply to the room on the stack ventilation effectiveness. The effectiveness of stack ventilation was tested in a classroom during winter months. The improvement of the effectiveness of stack ventilation was analyzed as a function of the number of open windows. An increase of 80% was obtained in the effectiveness of stack ventilation, when one window is opened. Lower increments were obtained when the number of windows opened were 2 or more. The maximum air change per hour during the tests was only 0.73 h^{-1}. Despite of the influence of outdoor conditions on stack ventilation, the maximum air change per hour was lower than the recommended number of exchanges in classrooms. As a result, recommended ventilation rates in school classes were not achieved when using stack ventilation.

1. Introduction

Stack ventilation is the basic and most common ventilation in Poland. This statement applies to public buildings as well as single-family and multi-family houses. Its function is to remove polluted air from the room [1-7], and thereby maintaining good indoor air quality [8].

Stack ventilation operates on the principle of pressure difference at the inlet and outlet of the channel. This is caused by the difference in air density, which depends on the temperature difference. The condition of proper operation of stack ventilation is to bring outdoor air into the room. According to PN-83/B-03430 [9], "stack ventilation ducts can only be individual, the cross-section should be determined at an outdoor temperature of + 12°C". At lower outside temperature, and therefore with a larger temperature difference, there should be better chimney draft. This will cause a larger stream of air removed from the room [2,6]. The larger stream of air removed is not improving indoor air quality. A larger stream of removed air can lead to the cooling of air in the room and this will definitely lead to energy losses in the building [3,5,7,10-16].

Research on the effectiveness of natural ventilation, and above all the effectiveness of periodic airing of rooms was carried out, inter alia, by Heiselberg & Perino [17]. They showed that with the appropriate frequency of short-term ventilation, the quality of indoor air and the thermal comfort felt in accordance

IOP Conf. Series: Materials Science and Engineering **809** (2020) 012008 doi:10.1088/1757-899X/809/1/012008

with Directive 2002/91/EC [18] are improved. Stack duct ventilation will work properly if we bring outside air into the room.

Due to the fact that stack ventilation is the most unreliable installation [5]. Many authors are considering improving its effectiveness [17,19].

The studies on the effectiveness of stack ventilation in kindergarten in Białystok before cleaning the channels and after their purification showed an improvement in the effectiveness of stack ventilation by a minimum of 10%, a maximum of 100%. However, the supply of air through a permanent unsealing of windows, increased the effectiveness of stack ventilation by 30% [20,21].

The aim of the study was to determine the effectiveness of stack ventilation in one classroom and determine the impact of additional air supply to the room in the stack ventilation effectiveness.

2. Methodology and research

The research on the effectiveness of ventilation in the didactic room (room for technical classes) was carried out in December. The room volume is 168.58 m^3 (Figure. 1). The room is located on the lowest floor of the building. It is not tight room because there was a 2 cm gap under the door. The length of the ventilation duct is 12 m.

Figure. 1. The room where the measurements were made.

Measurements of the inlet air velocity through the windows and the exhaust air velocity through grilles were made with an anemometer probe connected to the TESTO 435-4 measuring instrument. Measurements of the dimensions of grilles, dimensions of the window aperture, volume of the room were made with the Bosch 50 laser rangefinder. The parameters and accuracy of the measuring instruments are given in literature [22,23].

The volumetric exhaust air flow rate, $\dot{V}$ was calculated by using the equation of continuity (eq. 1):

$$\dot{V} = A\ V \quad (1)$$

Where:

A - cross-sectional area of the air stream outflow from the room (cross-sectional area of the stack ventilation grid) [m^2]; in the tested room is 0.0121 m^2.

V - velocity of air outflow from the room [m/s].

During experimental tests, outdoor air temperature was 0°C, and the indoor air temperature was 22°C. The results of the measurements are shown in Table 1.

The exhaust air flow rate was higher when the number of open windows increased, as shown in Figure. 2. An important increment was obtained in the exhaust air flow rate when one window was opened.

It should be stated that in the case when there was no additional air supply, a low value of exhaust air flow rate, 19.6 m^3/h, was obtained by stack ventilation. In this case, air flow was flowing through the gap under the door and through leaks in the building woodwork and through the structure of the wall material [7].

Table 1. Measured and calculated air parameters.

Parameter / variant						
Number of open windows	-	0	1	2	3	4
Inlet air flow rate through window	m^3/h	0	131	262	393	524
Mean air velocity through grid 1	m/s	0.25	1.13	1.35	1.35	1.37
Mean air velocity through grid 2	m/s	0.20	1.2	1.39	1.45	1.45
Mean air velocity through grilles	m/s	0.225	1.165	1.37	1.4	1.41
Exhaust air flow rate through grilles	m^3/h	19.60	101.49	119.35	121.97	122.84
Air changes per hour	h^{-1}	0.12	0.60	0.71	0.72	0.73
Increased ventilation efficiency	%	-	80.69	14.96	2.14	0.71

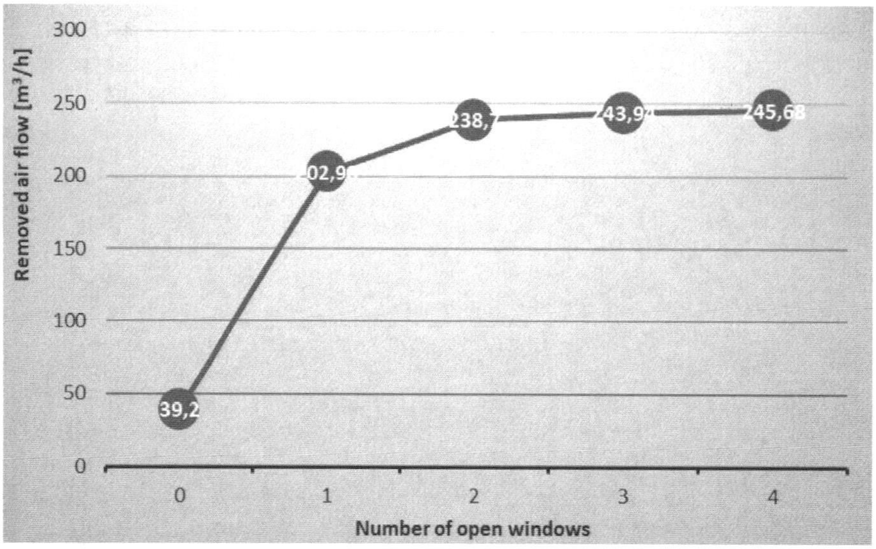

Figure. 2. Exhaust airflow rate and number of opened windows.

An increase of 80% was obtained in the effectiveness of stack ventilation, when one window was opened, as shown in Figure. 2 and Table 1. The opening of a second window resulted in an increase of the stream of air supplied to the room by 100% to the value of 262 m^3/h and an increase in ventilation efficiency of 14.96%. Opening the third or fourth window leads to an increment of 2.14% and a 0.71% in the effectiveness of stack ventilation respectively.

An effectiveness of stack ventilation of 0.12 h^{-1} was obtained in the case with closed windows (in the absence of outdoor air supply). This value is in the typical range than other case studies where the effectiveness of stack ventilation should range from 0 to 0.2 h^{-1} [24]. In the case study with one opened window, a value of 0.6 h^{-1} of air exchange was obtained. This value is in the range than other reported in the literature where the number of exchanges with tilted windows should range from 0.3 to 3 h^{-1} [24].

Nevertheless, the recommended number of exchanges in classrooms should be between 3 and 5 h^{-1} [24]. However, the maximum air change per hour during the tests was only 0.73 h^{-1}.

In summary, further increasing the incoming air flow is not recommended. It will only cool the room. In this particular room, one should consider: whether to bring the room to the air stream through one window, which will improve the efficiency of the exchange by more than 80%, or by two, which in total will result in a 95% improvement in the effectiveness of stack ventilation.

3. Conclusions

Experimental tests were made to analyze the effectiveness of stack ventilation in a classroom. The experimental tests analyzed the influence of opening windows in the effectiveness of stack ventilation.

A low value of exhaust air flow rate was obtained by stack ventilation in the case with closed windows. An increase of 80% was obtained in the effectiveness of stack ventilation, when one window was opened. Lower increments were obtained when the number of windows opened were 2 or more.

The maximum air changes per hour, 0.73 h^{-1}, was obtained with four windows opened. However the obtained value is lower than the recommended [24] number of air changes per hour in classrooms. So, despite outdoor conditions influence on air exchange by means of stack ventilation, insufficient ventilation air stream recommended in school classes was not achieved [24]. The described method of supporting stack ventilation can be used, during short periods of time in similar rooms, when the stack ventilation is the only ventilation system.

Acknowledgment
The study was carried out within the framework of the statutory project financed by the Polish Ministry of Science. The work was financed from the project BUT InterAcademicPartnerships PPI/APM/2018/1/00033/DEC/1.

4. References

[1] A. Acred, and G. R. Hunt, Build. Environ. vol. 72 (2014), p. 44.
[2] K. Arendt, M. Krzaczek, and J. Tejchman, Build. Simul. vol. 10(2) (2017), p. 229.
[3] M. Ferrucci and M. Brocato, Build. Serv. Eng. Res. T. vol. 40(1) (2019), p. 109.
[4] T. Gaczoł, J. Civil Eng., Environ. Arch. vol. 62(2) (2015), p. 81.
[5] G. C. da Graça and P. Linden, Build. Environ. vol. 107 (2016), p. 263.
[6] M. Krzaczek, J. Florczuk, and J. Tejchman, Energ. Buildings vol. 103 (2015), p. 48.
[7] R. Tamašauskas, J. Šadauskiene, P. Bruzgeviˇcius and D. A. Krawczyk, Energies vol. 12(11) (2019), p. 2145.
[8] P. Kapalo, H. Klymenko, V. Zhelykh and M. Adamski, Lecture Notes in Civil Engineering vol. 47 (2020) p. 168-173.
[9] PN-83/B-03430.
[10] V. V. Pudikov, N. B. Litvinova and P. A. Grushkovsky, in *2018 International Multi-Conference on Industrial Engineering and Modern Technologies (FarEastCon)*, Vladivostok, Russia, p. 18393101 (2018). 10.1109/FarEastCon.2018.8602840.
[11] A. M. Rodrigues, M. Santos, M. G. Gomes and R. Duarte, Buildings vol. 9(5) (2019), p.123.
[12] D. A. Krawczyk, Energy Procedia vol. 95 (2016), pp. 216.
[13] D. A. Krawczyk and T. J. Teleszewski, Energies vol. 12(6) (2019), p. 1012.
[14] A. M. Marina Domingo, J. M. Rey-Hernández, J. F. San José Alonso, R. Mata Crespo and F.J. Rey Martínez, Energies vol. 11(10) (2018), p. 2826.
[15] A. Mikola, R. Simson and J. Kurnitski, Energies vol. 12(13) (2019), p. 2633.
[16] A. Tejero-González, D. A. Krawczyk, J. R. Martín-Sanz García, F. J. Rey-Martínez and E. Velasco-Gómez, Energies vol. 12(15) (2019), p. 3033.
[17] P. Heiselberg, M. Perino, Indoor Air vol. 20 (2010), p. 126.
[18] Directive 2002/91/EC of European Parliament and the Council of 16 December 2002 on the Energy Performance of the Buildings.
[19] K. Gładyszewska-Fiedoruk, V. Zhelykh and A. Pushchinskyi, Energies vol. 12(15) (2019), p. 2845.
[20] T. Teleszewski, E3S Web Conf., 44, (2018), 00177.
[21] A. Asif, M. Zeeshan and M. Jahanzaib, Atmos. Pollut. Res. Vol. 10(2) (2019), p. 531.
[22] K. Gładyszewska-Fiedoruk, E3S Web Conf., 49, (2018), 00032.
[23] K. Gładyszewska-Fiedoruk and M. Nieciecki, Energy Procedia vol. 95 (2016), p. 132.
[24] H. Recknagel, E. Sprenger, and E.-R. Schramek: *Taschenbuch für Heizung + Klimatechnik 07/08, einschließlich Warmwasser- und Kältetechnik Gebundene Ausgabe*, 73 ed. Vulkan-Verlag GmbH, Dec. 2006.

IOP Conf. Series: Materials Science and Engineering **809** (2020) 012009 doi:10.1088/1757-899X/809/1/012009

Examples of the HVAC Systems' Modernization in the Existing Schools and Kindergartens

Dorota Anna Krawczyk[1, a], Antonio Rodero[2,b] and Tomasz Janusz Teleszewski[1,c]

[1]Bialystok University of Technology, Wiejska 45A 15351 Bialystok, Poland
[2]University of Cordoba, Campus de Rabanales 14071 Cordoba, Spain

[a]d.krawczyk@pb.edu.pl, [b]a.rodero@uco.es, [c]t.teleszewski@pb.edu.pl

Abstract. The possibility to reduce energy consumption in a group of 30 public, education buildings located in Poland, by improvements in the HVAC systems, was analyzed. The study group included both: primary and secondary schools as well as kindergartens. 14% of evaluated heating systems were found as having a proper efficiency and technical quality, while in a case of 86% changes regarding heat regulation, distribution or generation were recommended. Theoretical reduction of energy for heating could reach a level of 5-18% in buildings with water pipes systems.

1. Introduction

According to the Energy Performance of Buildings Directive, all new buildings in EU countries must be nearly zero-energy buildings by 31 December 2020 and a huge attention is paid on the public buildings. Majority of existing buildings were constructed prior to introducing any formal energy performance requirements, as a result of which the quality of the building stock is considerably below that which can be achieved today. Therefore, they need appropriate modernization, that is improvement of existing building technical features of a building which should lead first of all to reduction in energy demand. In Poland, nearly 37% of total energy in non-residential buildings is consumed by heating, ventilation and air conditioning (HVAC) [1].

In Poland a huge part of energy is used for heating of buildings, in contrast to countries located in south Europe, like for example Spain, or Italy where much more energy usage is connected with cooling [2]. Energy situation of new buildings differs significantly from existing, old ones. For example the kindergarten that was built in 2014 as an energy efficient kindergarten, has a heating demand of approximately 35-40 [kWh/(m^2·year)]. As described in [3] the building is heated with floor heating supplied by low-temperature natural gas boilers.

As shown in previous studies [1,4,5], the heating energy consumption factor in Polish existing schools and kindergartens was found between 133.5 [kWh/(m^2·year)] and 261.1 [kWh/(m^2·year)]. Thus, there is a possibility to apply solutions that can reduce energy consumption, such as retrofitting of envelope, replacing of windows by new ones with a low U value, as well as modernization of the HVAC system. Efficiency of the ventilation system on the overall energy consumption and the indoor air quality (IAQ) was described in [6,7]. Issues of efficiency of distribution network was discussed in [8].

In this paper several buildings of schools and kindergartens located in Bialystok, Poland were selected and possibility to improve HVAC systems was analysed, as well as a range of improvements applied last years was shown.

2. Description of buildings

The group of 30 kindergartens and schools located in Podlaskie Province, Poland was analysed. Buildings were constructed between 1919 and 1988, thus structures of walls and roofs were design to meet regulations regarding heat transfer coefficients that were effective these years. During the retrofitting process, to reduce heat losses through heat transfer, in case of most of buildings (87%) walls and roofs were isolated and external windows, doors replaces by new ones. These improvements allowed to reduce useful energy consumption in the buildings.

All buildings had heating systems, while there were no cooling systems in them, except few offices used temporary by buildings' administration staff.

In all buildings natural ventilation was used in rooms for students and pupils, while a mechanical ventilation was used only in kitchens, if any.

3. Description of the HVAC systems and thermal comfort

Majority (29 of 30) of buildings had a water heating system with column and/or panel radiators, that is the most often used in Poland. Installations were supplied from the Central Town CHP (Combined Heat and Power) Centre. Heating systems were used continuously all months of the heating season, that lasts from September to April/May depending on the year.

The last building (the kindergarten located in a rural area dedicated for children living in villages in the nearest surrounding) was equipped in electric radiators, and according to information from a building owner the heating installation was used only in a kindergarten working hours (average 8 hours per day from Monday to Friday).

Installations with a water heating system were modernized during exploitation periods, whereas the range of improvements was different, as shown in Figure. 1.

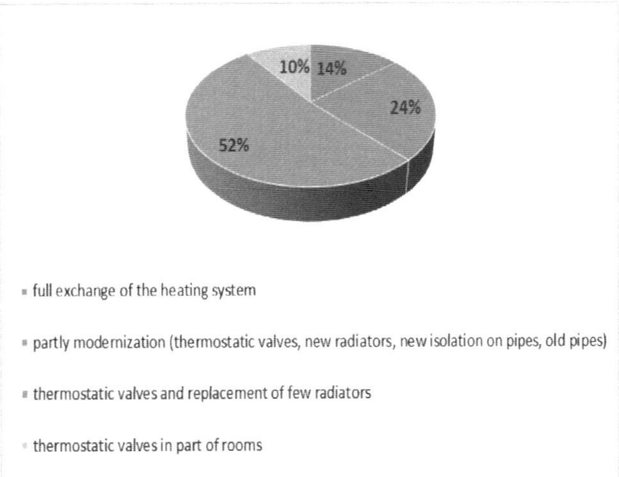

Figure 1. The range of improvements in the heating system.

4. Indoor thermal comfort

Design temperature in Poland for classrooms and offices is set down as 20°C. In the analysed buildings the indoor temperature during normal usage of classes was found between 16.8°C and 21.2°C.

The lowest value was recorded in the kindergarten with electrical heaters in the morning. In other buildings (with a water heating system) temperature didn't fall away below 18.8°C.

It is very important to maintain a proper indoor temperature, as shown in many previous studies it influences people sensation and students aptitude for learning [9-13].

5. Evaluation of heating system efficiency and analyse of possible improvements

Firstly, buildings with a water heating system were considered.
According to Eq.1 the total efficiency of heating system depends of 4 factors.

$$\mu_{tot}= \mu_r\, \mu_g\, \mu_d\, \mu_a\ . \qquad (1)$$

where:
μ_{tot} is total seasonal efficiency of the heating system,
μ_r is the average seasonal efficiency of heat regulation and use in heated space, [−],
μ_g is the average seasonal efficiency of heat production from an energy carrier or the energy supplied to a heat source, [−],
μ_d is the average seasonal efficiency of heat transmission from a heat source to heated space, [−],
μ_a is the average seasonal efficiency of heat accumulation in the inductance components of a heating system, [−],
Thus, the higher total efficiency of HVAC system is, the lower final energy consumption we obtain Eq.2:

$$Q_k = \frac{Q_{nd}}{\mu_{tot}}\ . \qquad (2)$$

where:
Q_k – the annual demand for final energy supplied to a building or its part for the central heating system, [kWh/year],
Q_{nd} – the annual energy demand for heating space, [kWh/year].
Only in a case of 14% analysed buildings (Figure.1) no improvements are recommended, while for 86% of systems it is possible to improve their efficiency. Most of all it is recommended to install thermal valves in rooms where they are missing, to ensure stable indoor conditions and prevent from overheating. Moreover in most cases old radiators should be replaced by new panel ones with a low capacity, as well as old pipes and isolation. In all cases buildings were supplied by heat centres from district heating and it was not necessary to improve the energy source. It is worthy to note that is necessary also to consider possibility to improvements in regulation system even in buildings were thermostatic valves were installed, but they were old types and in many cases not working properly. As shown in [14] in another part of Poland actions to replace old thermostats with Danfoss radiator sensors and thermostatic valves in buildings managed by the city, e.g. in the schools, municipal facilities, city offices, and other socially important facilities, were taken. It is predicted that thanks to better temperature control in individual rooms, about 23 percent of energy could be saved. Moreover it would result in better indoor thermal comfort and would help to avoid situations when part of rooms of overheated while in the others indoor temperature is too low to meet requirements of students and teachers.

Secondly, for the kindergarten with electric radiators, due to a fact that the building used only in mornings and it is located far from a district heating network, it is highly recommended to install an air heat pump.

6. Summary

Improvements in water heating systems could result in 5-18% reduction of energy consumption while replacement of electrical radiators by the air heat pump would increase the system efficiency even 2-4 times, depending of its COP (coefficient of performance).

Findings are consistent with conclusions presented in [15], where analysis of 4 schools showed that in a case of buildings built before 1945, it was possible to improve the efficiency of the installation by about 17%, whereas in buildings built after 1970 - by about 5%.

Acknowledgments
The study was carried out within the framework of Polish Ministry of Higher Education and Science grant WZ/WBiIS/9/2019. The work was financed from the project BUT InterAcademic Partnerships (PPI/APM/2018/1/00033/DEC/1) NAWA.

7. References

[1] D.A.Krawczyk, B.Sadowska: *Buildings 2020+. Constructions, materials and installations.* edited by D.A.Krawczyk, Printing House of Bialystok Univesity of Technology Bialystok – Cordoba – Vilnius (2019)

[2] D.A.Krawczyk, B.Wądołowska, A.Rodero: E3S Web of Conferences Vol.44, (2018)

[3] Information on http://www.renew-school.eu/wp-content/uploads/2014/10/16_Siemianowice_folder.pdf

[4] C.A. Balaras, K. Droutusa, E.Dascalaki, S. Kontoyiannidis: Energy and Build. Vol. 37 (2005), p. 429

[5] D.A.Krawczyk: Energy Procedia Vol. 95 (2016) p. 216

[6] D.A.Krawczyk, K.Gladyszewska-Fiedoruk: Rynek Energii, Vol.5 (2013), p. 63

[7] K. Gładyszewska-Fiedoruk, V. Zhelykh, A. Pushchinskyi; Energies Vol.12 (15), (2019), p.2845

[8] D.A. Krawczyk, T.J. Teleszewski: ENERGIES Vol. 12 (11), (2019)

[9] M.Á. Campano, S. Domínguez-Amarillo, J. Fernández-Agüera, J.J. Sendra: Sustainability Vol. 11(14) (2019), p. 3948

[10] M.Talejko: Environment (2015), p.194

[11] T. Kapwata, M.T. Gebreslasie, A.Mathee, C.Y.: Wright Int J Environ Res Public Health, 15(5), (2018) p. 952.

[12] Z.Wang, Y.Ji, X. Su: Building and Environment Vol.133 (2018)

[13] A.Lis, Z. Vranayova : CzOTO, Vol.1 (2019), p. 316-323

[14] Information on https://www.danfoss.com/en/service-and-support/case-studies/cf/first-step-towards-smart-city-in-poland/

[15] A. Alsabry: Energy and Buildings Vol.130 (2016), p.811.

IOP Conf. Series: Materials Science and Engineering **809** (2020) 012010 doi:10.1088/1757-899X/809/1/012010

Recognition of Nonlinear Hysteretic Behavior by Neural Network using Deep Learning

Taiji Mazda[1, a], **Yukihide Kajita**[1,b], **Takashi Akedo**[1,c] **and Tsubasa Hazama**[1,d]

[1]Department of Civil Engineering Kyushu University, 744 Motooka Nishi-ku Fukuoka, Japan

[a]mazda@doc.kyushu-u.ac.jp, [b]ykajita@doc.kyushu-u.ac.jp,[c]akedo@doc.kyushu-u.ac.jp , [d]hazama@doc.kyushu-u.ac.jp

Abstract. In Japan, large earthquakes have caused many damages to structures. Cracks in the reinforced concrete and plasticization of the steel bar and steel components resulted in increased damage. In the seismic design, the dynamic response analysis is carried out by using a mathematical model that appropriately evaluates the nonlinearity of the material and the components and the seismic performance is confirmed by performing a dynamic response analysis. However, when applying new materials and components, much time and much effort are required to select a mathematical model. In this study, we focused on the high pattern recognition capability of the neural network. We attempt to directly model using a neural network without replacing the complex nonlinear hysteretic behavior using the mathematical model. By improvement of learning data and introduction of deep learning, it was confirmed that the recognition ability for nonlinear hysteretic behavior of neural network was greatly improved, and its applicability as a numerical operation subroutine for time history response analysis was drastically improved.

1. Introduction

In the seismic performance evaluation of the highway bridges with nonlinear dynamic response analysis, some modeling is required to evaluate how they affect the seismic response when new materials and newly developed components are applied to the bridges. Usually, various static loading tests, dynamic loading tests, and in some cases, shaking table tests are conducted. It is possible to confirm the complex nonlinear load-displacement relationships, the energy absorption performance, and the damping characteristics from the results of these tests. Based on these characteristics, a mathematical model capable of expressing complex load-displacement relations is proposed and incorporated into the analysis program. The mathematical models obtained in this procedure can reasonably express the single nonlinear hysteresis behavior of the materials and members to be evaluated, but it is usually necessary to make a drastic correction when applying to the expression of the nonlinear hysteretic behavior of different materials and components. J. Ghaboussi et al. [1] conducted a pioneering study to apply neural networks to the recognition of the nonlinear stress-strain relationship of concrete. K. Yamamoto [2] also applied the neural network to the nonlinear hysteresis recognition of the Ramberg- Osgood model. J.D. Wei [3] applied the neural network to the construction of hysteresis, too. In this study, we focused on the high pattern recognition capability of the neural network. We attempt to model directly using a neural network without replacing the complex nonlinear hysteretic behavior using the mathematical model.

2. Construction of Neural Networks

2.1 Neuron Model.
Neuron model shown in Figure. 1 can simulate the information processing of brain nervous system industrially. This information processing organization is called a node in following. The characteristic of this node has multiple inputs and single output. One node receives the input signals from n nodes. These input values are x1, x2, … , xn. W shows strength of connection between each node, and is called the weight of connection. b is the bias input to restrain or accelerate the output of node. Input signals x1, x2, … , xn are amplified or reduced by the weight of connection, and are summed up as internal potential x. X is transformed to output value Y through fReLU (x) which is called activation function. ReLU shown in formula (1) was used as activation function in this study. And identity function was used only for output layer.

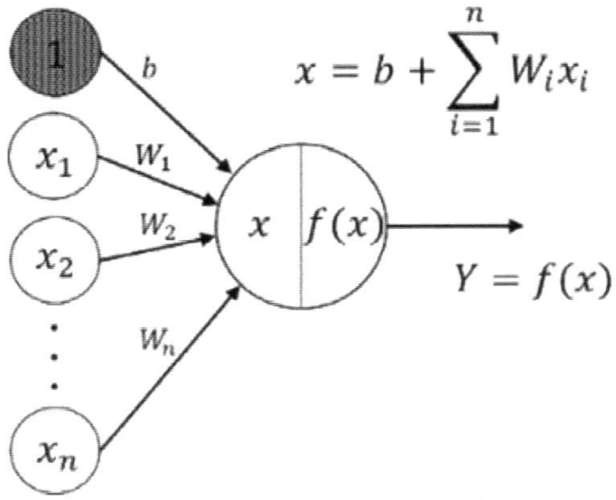

Figure 1. Neuron model

$$f_{ReLU}(x) = \max(0, x) \qquad (1)$$

2.2 The Hierarchical Neural Network.
Hierarchical neural network was adopted in this study. Information is transmitted toward the output layer from the input layer in one-way direction on the network. Neural network transmits the signals from the input layer to the output layer by way of the hidden layers. When an error between these output signals and the data for learning occurred, the weights of connection are modified to reduce this error. This modification process is equivalent to learning of neural network. Modification proceeds to minimize the error toward the input layer from the output layer in learning algorithm. The hierarchical neural network used in this study is shown in Figure. 2. In this study, we set 11 nodes in the input layer with reference to the results of the previous study. The hidden layer was made into 6 layers and the number of nodes was set to 100.The output layer consists of 1 nodes and outputs the tangential stiffness at the present time.

2.3 Selection of Information in Input Layer.
The information that can evaluate complex hysteretic behavior like a Ramberg-Osgood model need to be selected adequately for the input layer of hierarchical neural network. It is to be desired that selected information can be taken easily from loading tests as the data for learning of the network. The information in input layer were selected as follows referring the past studies[4-8]. Hysteretic curve including non-linear behavior like a Ramberg-Osgood model depends upon magnitude of displacement and load. Hysteretic curve forms the most outer loop renewing maximum experienced point when loading direction changes. So, maximum displacement, maximum load, minimum displacement and minimum load were selected as an information in input layer. And when maximum displacement isn't renewed, hysteretic curve goes toward for opposite maximum experienced point. Load and displacement at latest turning point are important as an information in input layer. Displacement and load were considered at maximum

IOP Conf. Series: Materials Science and Engineering **809** (2020) 012010 doi:10.1088/1757-899X/809/1/012010

experienced point and the latest turning point individually. Increases of previous displacement, increases of previous load were added as information to consider the transformation of stiffness. Previous displacement and previous load were added to consider the continuity hysteretic curve. Above ten information and present displacement were defined as inputs for the input layer. The output layer has one node of present tangential stiffness.

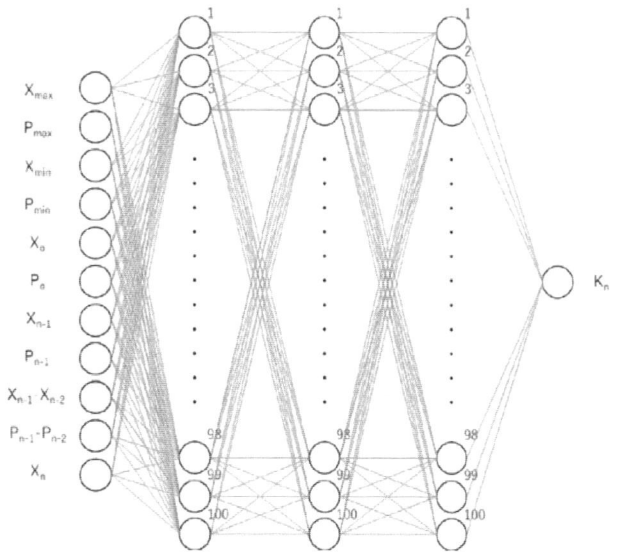

X_{max} : maximum experienced displacement
X_{min} : minimum experienced displacement
P_{min} : minimum experienced load
X_o : the latest displacement at turning point
P_o : the latest load at turning point
X_{n-1} : previous displacement
P_{n-1} : previous load
X_{n-1}-X_{n-2} : increase of displacement
P_{n-1} - P_{n-2} : increase of load
X_n : present load
K_n : present tangent stiffness

Figure 2. Hierarchical neural network

2.4 Recognition of Non-linear Hysteretic Behavior.

Forced displacement of increased and decreased sinusoidal wave was given to non-linear spring of Ramberg-Osgood model. The relationship between road and displacement was adopted as a data for learning of the neural network. Forced displacement was formed with continuously and gradually increased and decreased sinusoidal wave which the period is 0.25 second with an amplitude of 20%, 40%, 60%, 80%, 100% of the maximum one. Maximum amplitude was 5cm. Figure. 3 shows time history of forced displacement. Using this data, the learning for neural network was conducted. The repetition of learning of neural network was one million times. Figure. 4 shows the comparison of hysteretic curves that was estimated by the network after learning and the hysteretic curve of the data for learning in Ramberg-Osgood model. In previous studies, only learning data such as Figure. 3 were prepared, but this time, the superimposed sinusoidal wave with small amplitude and high frequency sinusoidal wave was prepared. Figure. 5 shows superimposed sinusoidal wave. The data for learning were prepared using this superimposed forced displacement. By using this forced displacement, the load displacement relationship, which is closer to the behavior in the actual earthquake, is reapplied, and the construction of high quality learning data is expected.

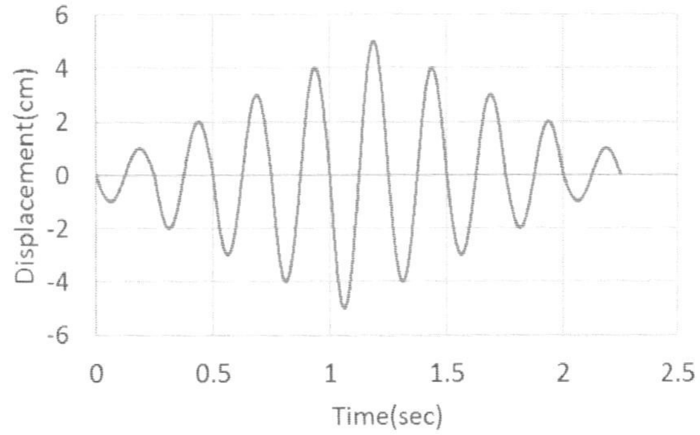

Figure 3. Time history of sinusoidal forced displacement

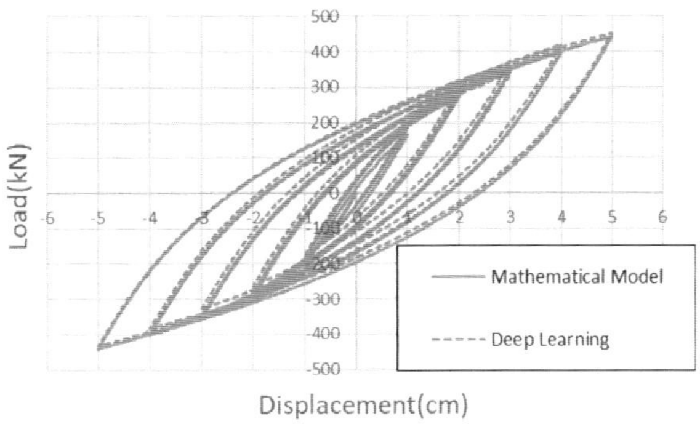

Figure 4. Comparison of hysteretic curves

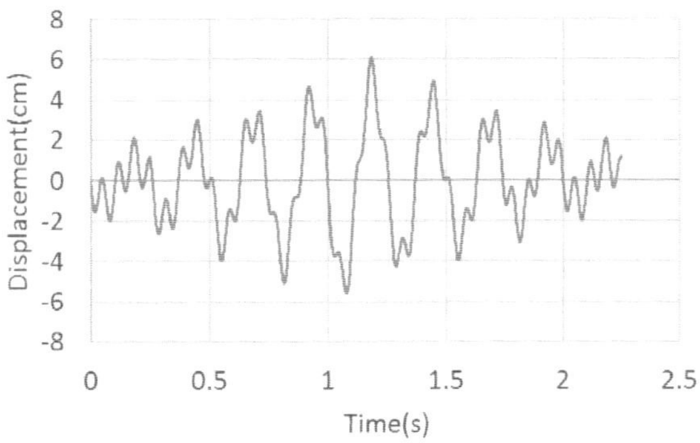

Figure 5. Time history of superimposed forced displacement

3. Dynamic Response Analysis using Neural Network

3.1 Input Earthquake Motion and Analytical Model.

The input earthquake motion used for the analysis is shown in Figure. 6. The earthquake ground motion is a record of Kawayo point in the South Aso village, which was observed at the Kumamoto Earthquake in 2016. The main part of the ground motion was adjusted so that the response displacement was not higher than the maximum amplitude of the learning data of 5 cm. Figure. 7 shows an analytical model used for dynamic response analysis using a neural network. The neural network, which was trained so that the tangential stiffness at any displacement can be output, was incorporated as a numerical arithmetic subroutine representing the springs of the analytical model of the single degree-of-freedom system.

IOP Conf. Series: Materials Science and Engineering **809** (2020) 012010 doi:10.1088/1757-899X/809/1/012010

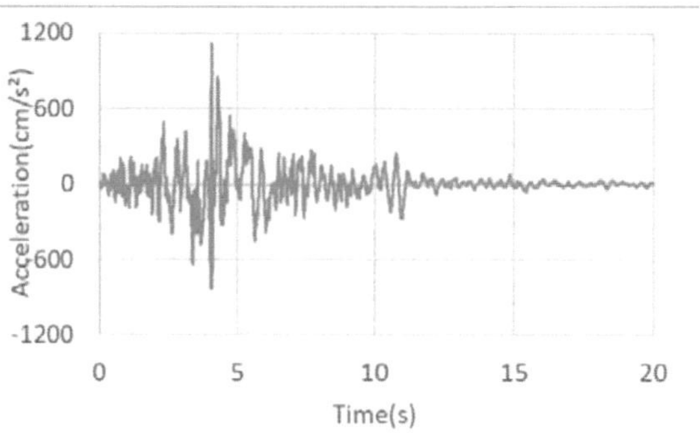

Figure 6. Input earthquake motion

3.2 Dynamic Response Analysis.

Figure. 8 shows a comparison of the time history obtained by using a neural network learned with sinusoidal forced displacement and the time history of the response displacement obtained by using the mathematical model. Comparing the analytical results, the results obtained by using neural networks in the first half of the analysis time showed a relatively good agreement in the amplitude and phase with the results obtained by using the mathematical model. However, in the latter half of the analysis time, there are differences of the amplitude and phase due to the accumulation of the errors. Figure. 9 shows a comparison of the time history obtained by using a neural network learned with superimposed forced displacement and the time history of the response displacement obtained by using the mathematical model. From these results, obtained time history by using a neural network learned with superimposed forced displacement was roughly agreed with the time history by using the mathematical model in the phase. On the other hand, it was recognized that the amplitude was slightly larger or smaller in the first half of the analysis time as a result.

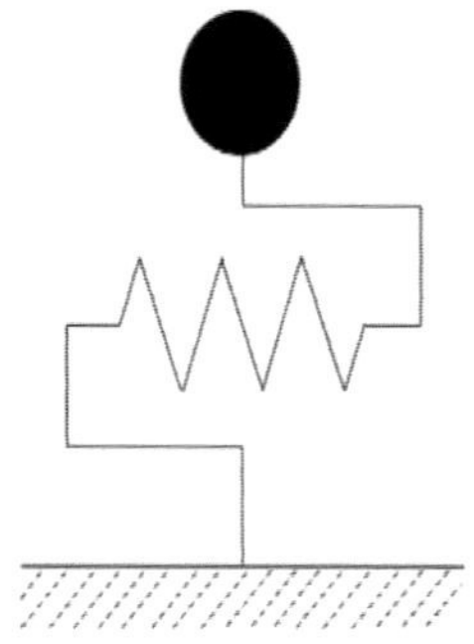

Figure 7. Single degree of freedom system

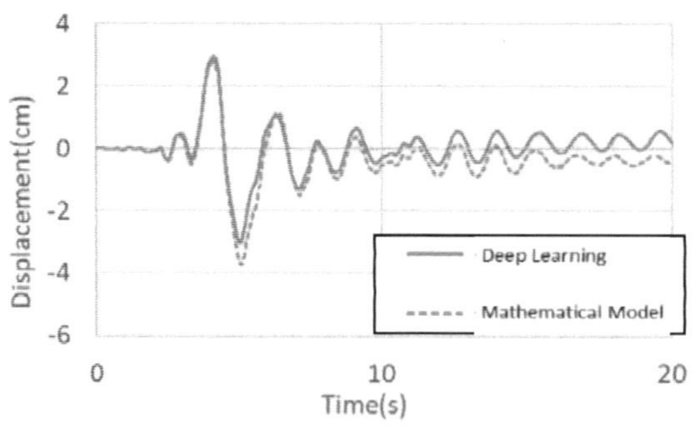

Figure 8. Comparison of time history
(leaned by sinusoidal forced displacement)

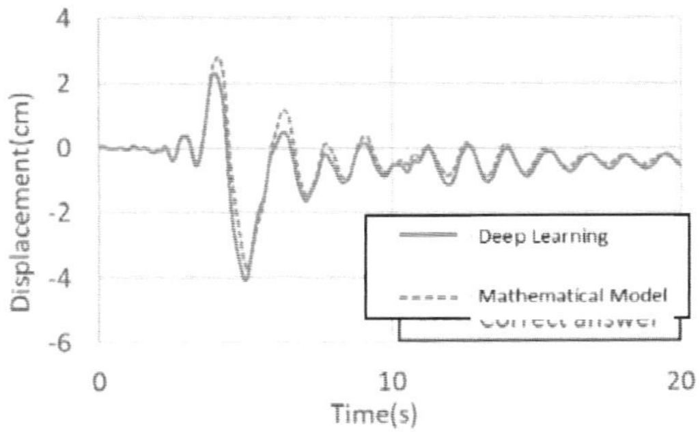

Figure 9. Comparison of time history
(leaned by superimposed forced displacement)

4. Conclusion

Function approximation ability of the neural network was evaluated through comparison between estimated response using neural network model and calculated response using mathematical models. From obtained results, the followings became clear.

Hierarchical neural network with deep hidden layer for recognition of non-linear hysteretic behaviors was constructed. Information of selected units in input layer can be taken easily from loading tests

An improvement proposal was made on how to provide the learning data. In addition to the sinusoidal wave used at the previous study, small amplitude sinusoidal wave with relatively short period was imposed for forced displacement. The load deformation relationship close to the actual dynamic behavior was realized.

The neural network, which has learned based on the improved learning data, reproduced the response equivalent to the response calculated using mathematical model in all duration time of the analysis.

Applicability of neural network as a subroutine to dynamic response analysis will be confirmed by dynamic loading test of new material and device in future.

Acknowledgement

The authors appreciate Kumamoto prefecture, Japan for providing the strong motion data.

References

[1] J. Ghaboussi, J.H. Garrett Jr and X. Wu, "Knowledge-Based Modeling of Material Behavior with Neural Networks", ASCE Vol.177, No.1, pp. 132-153, 1991.

[2] K. Yamamoto, "Modeling of Hysteretic Behavior with Neural Networks and its Application to Non-Linear Dynamic Analysis" Journal of Structural Engineering JSCE Vol.38A, 85-94, 1993.,

[3] J.D. Wei and C.T. Sun"Constructing Hysteretic Memory in Neural Networks", IEEE TRANSACTIONS ON SYSTEMS, MAN AND CYBERNETICS—PART B: CYBERNETICS, VOL. 30, NO. 4, pp. 601-609, 2000.

[4] T. Mazda, Y. Kabayama, T. Irie and T. Takayama, "A Study of Application of Neural Network to Non-Linear Dynamic Analysis", Journal of Structural Engineering, JSCE, Vol.42A, 635-644, 1996.

[5] T. Mazda, H. Otsuka and W. Yabuki, "Recognition of Hysteretic Behaviors of High Damping Rubber Bearing using Neural Network", ASME PVP-Vol.402-1, Seismic Engineering 2000, Vol.1, pp. 245-250, 2000.

[6] W. Yabuki, T. Mazda, and H. Otasuka, "Study of the nonlinear behavior of prestressed concrete girders by a neural network," Earthquake Resistant Engineering Structures, WIT Press,, pp.681-690, 2001.

[7] T. Mazda, H. Otsuka, W. Yabuki, and M. Tsuruta, "A study on generalized neural network system for recognizing nonlinear behavior of structures," Proceedings of the third international conference on engineering computational technology, No.85, Sep. 2002.

[8] T. Mazda, H. Otsuka, W. Yabuki, and K. Iwamoto, "Construction of Generalized Neural Network System for Recognizing Several Non-Linear Behaviors," ASME PVP-Vol.466, Seismic Engineering 2003, pp.143-151, 2003.

IOP Conf. Series: Materials Science and Engineering **809** (2020) 012011 doi:10.1088/1757-899X/809/1/012011

Effect of Number of Stories on the Seismic Soil Structure Interaction Performance of Midrise Frame Structures

Sahar A. Ismail [a], **Fouad K. Kaddah** [b] **and Wassim E. Raphael** [c]

Université Saint-Joseph de Beyrouth / Department of Civil & Environmental Engineering, Beyrouth, Lebanon

[a] sahar.ismail@net.usj.edu.lb , [b] fouad.kaddah@usj.edu.lb, [c] wassim.raphael@usj.edu.lb

Abstract. When an earthquakes hits, a state of emergency is declared that shakes lives and infrastructures and provokes aftershocks including fires and medical emergencies. The response of midrise concrete frame structures to a seismic excitation is related to the dynamic interaction between the structure, foundation and soil called soil structure interaction (SSI). To study the effect of structures' number of stories on the seismic behavior of midrise concrete frame buildings including soil structure interaction effects, 5 to 15 stories structures were simulated by performing 3D time history nonlinear finite element simulations using Abaqus under the influence of El-Centro (1940) and Northridge (1994) earthquakes. The models consisted of frame structures supported by raft foundation and rested on silty sandy soil block. The results presented in terms of story lateral deflection, inter-storey drift, shear force, foundation rocking and response spectrum show that structures' number of stories significantly affect SSI effects and increasing story number can end in structures' collapse in some cases.

Notation List: c= soil apparent cohesion value[kPa], γ_d=dry unit weight[kN/m^3], ρ =density[kg/m^3], Φ= internal friction angle[°],Ψ=dilation angle[°], E=Young Modulus[MPa], G= shear modulus[MPa], f'_c=concrete compressive strength[MPa] , e=void ratio, ν=poisson's ratio, PI=plasticity index[%], V_s=shear wave velocity[m/s], Υ_c= cyclic shear strain[%], A=acceleration[g], PGA=peak ground acceleration[g], S_a=spectral acceleration[g], ζ=damping ratio[%],d_i= deflections at i level[mm], h=the story height[m], α=mass damping factor, β=stiffness damping factor.

1. Introduction

When a 6.7 earthquake hit Northridge neighborhood in Los Angeles in 1994, at least 57 people died and $25 billion in damage were caused. This kind of strong earthquakes shook residents' lives and affected infrastructures forcing cities to declare states of emergencies. The response of midrise concrete frame structures founded on soft soils to strong shakings depends on the response of the soil medium beneath it having its own characteristics. Soil structure interaction (SSI) is the dynamic interaction problem between the structure, foundation and underlying soil or rock. SSI can either amplify or attenuate a seismic response depending on the soil response spectrum [1].

To improve the factor of safety, most seismic codes neglect the effects of SSI for structures founded on soft soils. This is due to the fact that SSI tend to increase structure's fundamental period, damping ratio and lateral displacement while they tend to increase or decrease structure's base shear depending on the soil spectral acceleration. Since 1970s, researchers have studied the effects of SSI on structures using analytical, experimental and numerical solutions [2, 3, 4, 5, 6, 7, 8, 9, 10, 11, 12, 13, 14, 15, 16, etc]. For the past 20 years, numerical solutions using finite element software SAP, Etabs, Ansys, Abaqus, etc and finite difference software like Flac are being used to model SSI problem using the direct method of analysis. Ref. [5] and Ref. [17] showed that SSI effects can only be neglected in soils having shear wave velocities below 600m/s. Moreover, Ref. [14] and Ref. [18] demonstrated that SSI effects can be neglected in cohesionless not cohesive soils. On the other hand, Ref. [6], Ref. [9], Ref. [12] and Ref. [15] showed that due to SSI effects, lateral displacement and thus inter-story drifts of flexible-structures are greater than those of fixed-based structures. While Ref. [12] proved that SSI effects get amplified with story height.

All these researches were performed on clayey soil (cohesive soils). Therefore, 3D nonlinear finite element simulations using Abaqus [19] were performed in this article to investigate the effect of number of stories on the seismic soil structure interaction performance of midrise concrete frame structures supported by raft foundation and rested on silty sandy soil under the influence of two strong ground motions: El-Centro (1940) and Northridge (1994). The results of flexible and fixed structures were in terms of story lateral displacement, inter-storey drifts, shear force, foundation rocking and response spectrum. The main objective of this paper is to understand the behavior of midrise concrete frame structures rested on silty sandy soil therefore, aid engineers in structures' design.

2. Finite element model

Buildings can be divided into low, medium and high rise depending on the number of stories (N): if N≤5: low-rise buildings, if 5≤N≤15: medium-rise buildings and if N≥ 15: high-rise buildings. The finite element software Abaqus 2017 [19] was used to study the effect of number of stories on the seismic performance of midrise concrete structures including soil structure interaction effects by varying N between 5 and 15. The results of the flexible models consisting of the structure, raft and soil block (SN-e-models with e referring to the raft thickness used in each case) were compared to the results of the fixed models consisting of the structure alone (SN-models) (Figure 1). All models were simulated using the direct method of analysis: the entire structure-raft-soil system was simulated at the same time and each model was hit at the bottom by El-Centro (1940) M_w=6.9, PGA=0.318g near field and Northridge (1994) M_w=6.7, PGA=0.843g far field earthquakes as presented in Figure 2. Noting that every time history model took around 60 hours to be completed using fast computational facilities at Université Saint-Joseph de Beyrouth.

Table 1 presents the model configuration details while Table 2 presents the properties of the materials used in this paper. The beams and columns were modeled using two-node linear beam B31 with 9400 elements while the slab was modeled using four-node double curved shell S4 with 100 elements. Moreover, 2.5 kN/m^2 uniformly distributed live and dead load was applied to the structures' floors. All frames were supported on a 20m width solid raft foundation modeled using eight-node linear brick, reduced-integration, hourglass control continuum solid elements C3D8R with 4800 elements. The structure-foundation system was then rested on a dense silty sandy soil solid block modeled also using C3D8R but with 96500 elements. While one-way infinite brick elements CIN3D8 elements were used to account for the absorbed energy from the unbounded soil domain in the horizontal directions. The beams and floor slabs were tied in the frame structure and a special tie procedure was performed between the columns and raft as well as between the raft and soil. In addition, embedded columns were inserted in the raft in flexible models.

Table 3 presents structures' natural frequencies calculated via linear perturbation procedure and Lancoz method available in Abaqus. Thus, the inelastic behavior of structural elements was modeled using elastic-perfectly plastic material while considering Rayleigh damping. 5% structural damping (ζ) was used to calculate the mass damping factor (α) and the stiffness damping factor (β) based on first and second mode frequencies (f_i and f_j in rad/s) as detailed in Table 3 where [20]:

$$\zeta = \frac{1}{2f_n}\alpha + \frac{f_n}{2}\beta \quad (1), \qquad \alpha = 2\zeta_i \frac{f_i f_j}{f_i + f_j} \quad (2) \text{ and } \qquad \beta = 2\zeta_i \frac{1}{f_i + f_j} \quad (3)$$

Moreover, the equivalent linear method was used to calculate the shear modulus (G) and damping ratio (ζ) of the silty sandy soil having a shear wave velocity of 63m/s [21,22,23]. A linear analysis using preliminary assumed values of G and ζ was performed. Then, the maximum cyclic shear strain recorded for each element in the soil model was implemented in the backbone curves given by Ref. [23] that relates G and ζ to the cyclic shear strain for different plasticity indices for cohesive soils. Therefore, the new G and ζ values obtained from these curves were implemented in the numerical model. This stage was repeated until no further change in G and ζ were obtained. Then, using first and second mode soil frequencies obtained also using Abaqus, Rayleigh damping coefficients were calculated for each earthquake as detailed in Table 2.

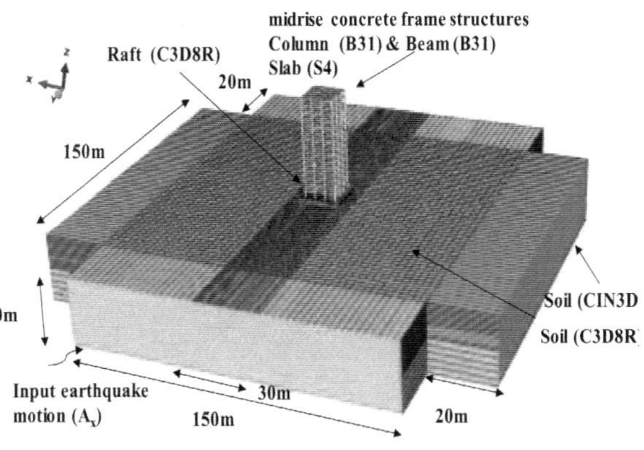

Figure 1. Geometry and model mesh distribution

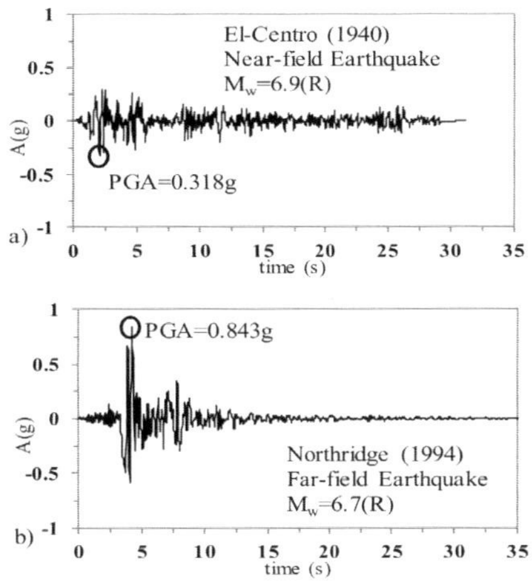

Figure 2. Acceleration with respect to time of a) El-Centro (1940) and b) Northridge (1994) earthquake

Table 1. Model configuration

Number of Bays	Story Height [m]	Bay Width [m]	Column [cm]	Beam [cm]	Slab [m]
3	3	5	50X50	50X50	5X5X0.25
Soil Dimensions			Raft Dimension		
Height [m]	Length [m]	Width [m]	Length [m]	Width [m]	Thickness [m]
30	150	150	20	20	0.5,0.7,1,1.2,1.5

Table 2. Material Properties

Concrete	E[GPa]	ν	ρ [kg/m^3]	f'_c [MPa]
	14	0.2	2400	20
	Silty Sandy Soil			
	γ_{max}[%]*	G/G_{max}*	ζ[%]*	α & β*
El-Centro (1940)	1.72	0.085	20.5	0.63, 0.066
Northridge (1994)	1.82	0.0769	21	0.65, 0.068
E [MPa]	ν	γ_d[kN/m^3]	e	* after [22]
20	0.3	18.98	0.4	
ϕ°	ψ°	c [kPa]	V_s[m/s^2]	PI [%]
35	10	50	63	15

Table 3. Structures' natural frequencies

Reference name	f1[Hz]	f2[Hz]	α*	β*
S5	1.079	3.679	0.52	0.003
S7	1.0651	3.251	0.5	0.004
S10	0.68276	2.075	0.32	0.006
S12	0.605	1.842	0.29	0.007
S15	0.473	1.445	0.22	0.008

* after [22]

3. Results and analysis

Previous studies compared SSI effects of flexible to fixed-based structures founded on cohesive soils [6,4,14,15,12,9,10, etc]. In this paper, the effect of number of stories on the seismic performance of midrise concrete frame structures while considering soil structure interaction effects was investigated by varying the stories between 5, 7, 10, 12 and 15 under the influence of El-Centro (1940) and Northridge (1994) earthquakes. Midrise soil structure interaction models: flexible-base models founded on 20m width raft foundation and supported by 150m X 150m wide and 30m height silty sandy soil block as shown in Figure 1 and Table 1 were simulated using the direct method of analysis. The soil boundary limits were chosen based on the recommendations of Ref. [24], Ref. [25] and Ref. [26]. Moreover, the results of flexible models, plotted in Figure 3 to 5 and presented in Table 4 and 5 in terms of story lateral deflection, inter-storey drifts, shear force, foundation rocking and response spectrum were then compared to those of fixed-base models.

3.1 Lateral deflections and inter-storey drifts

Inter-storey drift is a performance damage parameter that expresses the level of damage of a structure due to a seismic excitation. It is calculated as the difference between the lateral deflection of two stories divided by the story height:

$$drift = \left(d_i - d_{i+1}\right)/\,h \tag{4}$$

Performance levels are classified into 5 categories depending on the inter-storey drift value: if inter-storey drift < 0.2%: fully operational, < 0.5%: operational, < 1.5%: life safe, < 2.5%: near collapse and > 2.5%: collapse. Noting that life safe category is the acceptable category limit by most seismic codes [27, 28]. The lateral deflection at each story, presented when the maximum lateral deflection occurred at top of the structure regardless what time it occurred based on Ref. [26] and Ref. [29] was measured relative to the structure's base.

The results, plotted in Figures 4 and 5, showed that when structure's number of stories is greater than 10, lateral deflection of flexible models founded on silty sandy soils is in accordance with those founded on soft soils. In other words, lateral deflection of flexible models is greater than those of fixed-based models. This is in parallel with the results presented by Ref. [6], Ref. [9], Ref. [12] and Ref. [15]. However, lateral deflections of fixed-based lower stories structures: S5 and S7 were higher than those of flexible models. The ratio of flexible to fixed S15, S10 and S5 models varied between 1.9,1.33 and 0.5 and between 1.6, 1.7 and 0.9 under El-Centro (1940) and Northridge (1994) ground motions respectively. Therefore, the increase in lateral deflection from fixed to flexible models is related to the amount of energy the raft and soil absorb caused by the changes in the dynamic characteristics of the structure-soil-foundation system under a seismic excitation.

Moreover, lateral deflection at top of the structures increases with story number. For example, lateral deflection of flexible S5 to S15 models increased by a ratio of 8 and 6.7 at top of the structures and by 3.17 and 3 at the 5th level under El-Centro (1940) and Northridge (1994) earthquakes respectively. This increase in lateral deflection led to shifts in inter-storey drifts curves to higher and more dangerous categories. S10, S12 and S15 structures shifted from life safe category under El-Centro (1940) earthquake to near collapse and collapse categories under Northridge (1994) earthquake. Therefore, lateral deflections and thus inter-storey drift results were more affected by the stronger ground motion having higher PGA even though both earthquakes had similar magnitudes.

3.2 Shear force and Response Spectrum

To reflect the amount of energy the structure absorbs during an earthquake, shear force and response spectrum plots were generated. The maximum absolute shear force at each story level was calculated by summing up all shear forces produced by all columns in a given story. Using a Fortran code, the variation of pseudo-acceleration response spectrum plots (S_a) with different natural periods at 5% damping of horizontal accelerations formed at structures' base, at top of the soil profile (Free Field: FF) and of the input earthquake were created. Also, using MATLAB [30], Fourier transformation of the strong ground motions produced at top of the soil surface (FF) and at the base of the structure-foundation system were generated also at 5% damping. The results, illustrated in Figure 3 and 4, show that average level shear forces of flexible to fixed-base structures of S10, S12 and S15 structures were found greater than one whereas those of lower rise structures S5 and S7 structures were found lower than one under both earthquakes unlike the results obtained by [3, 6, 14, etc]. Hence, as the number of stories increase, structures' natural frequency tends to decrease and SSI effects get more significant. This is parallel with Ref. [31], Ref. [32] and Ref. [33] who showed that high frequencies structures are more affected by SSI effects than low frequencies structures. Response spectrum plots (Figure 5 and 6) present the seismic wave attenuation obtained from the input earthquake to the soil top surface (FF) caused by the soil type used: silty sandy soil. Noting that a wave amplification was obtained from FF to structure's base and structure-foundation system base similar to [25].

This amplification illustrates SSI effects. For example, S_a was attenuated from 0.94g at the ground level to 0.294g at FF then it was amplified to 0.50g at S15 flexible structure's base under El-Centro (1940) excitation. Thus, since level shear force tends to increase or decrease depending on the stiffness of the structure and the properties of the soil profile, the attenuation in the seismic waves caused lower frequencies structures to get more affected by SSI effects than higher frequencies structures.

Table 4. Average level shear ratios of flexible to fixed based cases

Number of stories	5	7	10	12	15
El-Centro (1940)	0.18	0.55	1.0	1.01	1.20
Northridge (1994)	0.65	0.26	1.11	0.97	0.98

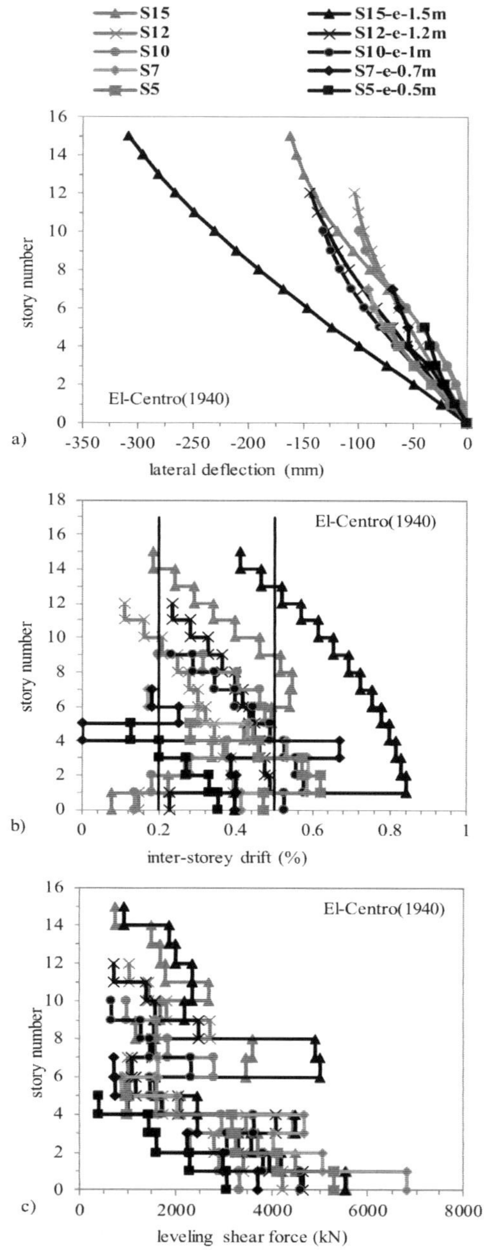

Figure 3. a) Lateral deflection, b) inter-storey drift and c) maximum shear force distribution of flexible and fixed-base models under El-Centro (1940) earthquake

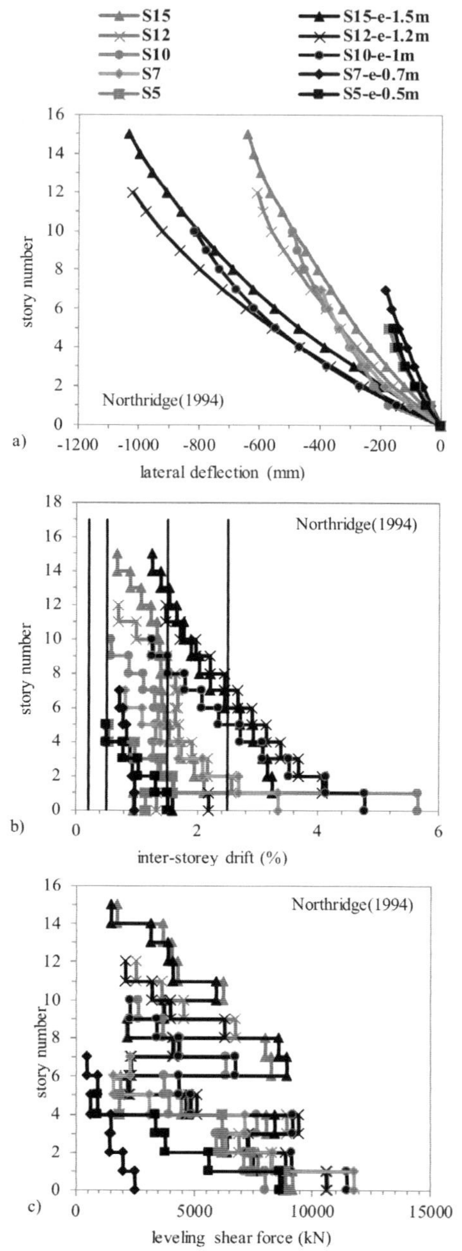

Figure 4. a) Lateral deflection, b) inter-storey drift and c) maximum shear force distribution of flexible and fixed-base models under Northridge (1994) earthquake

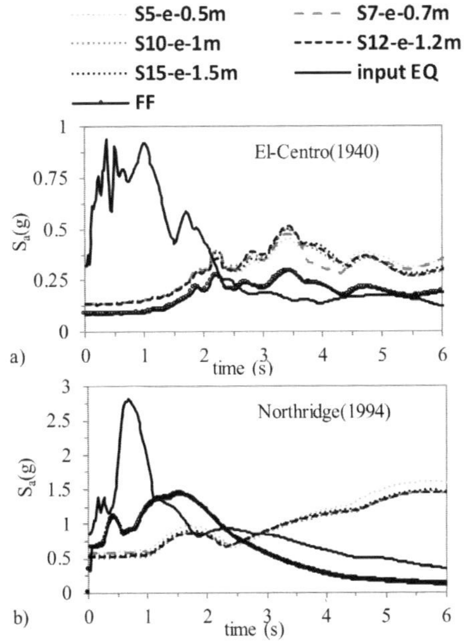

Figure 5. Acceleration response spectrum with 5% damping ratio for midrise frame structures under the influence of a) El-Centro (1940) and b) Northridge (1994) earthquakes

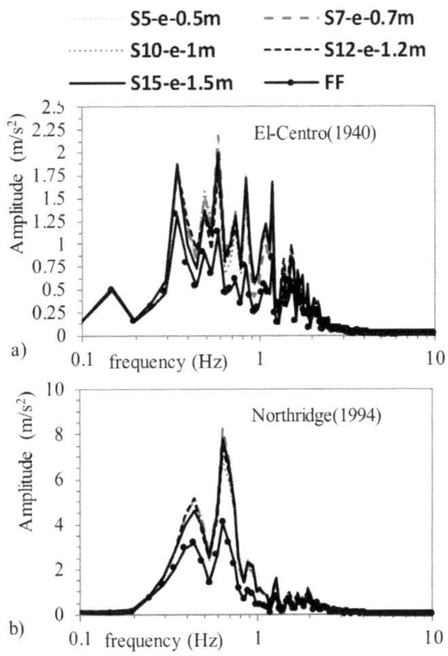

Figure 6. Variation of frequency content with the amplitude of midrise frame structures under the influence of a) El-Centro (1940) and b) Northridge (1994) earthquakes

3.3 Foundation Rocking

Relative structural displacement under a seismic excitation divided into structural and rocking components is caused by soil structure interaction effects [1]. Structural lateral deflection and inter-storey drifts depend on the amount of structural rocking experienced by a structure during a strong ground motion. Structural rocking is caused by inertial forces formed in a structural that cause compression on the first side and possible uplift on the other side of the foundation. Table 5 details the maximum foundation rocking experienced by the different simulated models. As shown in this table, the increase in structures' number of stories amplified soil structure interaction effects that led to the augmentation in foundation rocking under both earthquakes. S5 to S15 flexible structures maximum foundation rocking increased from 0.038° to 0.22° and from 0.11° to 0.44° under El-Centro (1940) and Northridge (1994) earthquakes respectively. To determine the amount of lateral displacement due to rocking from the foundation rocking angle, Ref. [34] and Ref. [35] relationship was used. For example, for S10 flexible structures under Northridge (1994) earthquake, the maximum foundation rocking angle was equal to 0.36° therefore, 36% of the maximum lateral deflection at top of S10-e-1m structure was due to rocking component and 64% was due to distortion component. In other words, the 816mm maximum lateral deflection at top of S10-e-1m structure was divided into 294mm due to rocking component and 522mm due to distortion component while the maximum lateral deflection at top of S10 structure was equal to 491mm and was due entirely to distortion component. As a conclusion, SSI effects increase foundation rocking angle reflected in the increase in the amount of structural and distortion component in flexible-based structures compared to fixed-based structures.

Table 5 Maximum foundation rocking angle [°]

Reference name	El-Centro (1940)	Northridge (1994)
S5-e-0.5m	0.04	0.11
S7-e-0.7m	0.054	0.24
S10-e-1m	0.11	0.36
S12-e-1.2m	0.16	0.42
S15-e-1.5m	0.22	0.44

4. Conclusions

The effect of number of stories on the behavior of midrise concrete seismic frame structures while including soil structure interaction effects was investigated by performing 3D nonlinear simulations using Abaqus. Fixed and flexible based structures (formed of the structure, raft and silty sandy soil block) were modeled and hit at the bottom by El-Centro (1940) and Northridge (1994) earthquakes. The results showed that low frequency structures got more affected by SSI effects than high frequency structures. This was reflected by lateral deflection, inter-storey drift, level shear force as well as response spectrum results. Moreover, the results showed that lateral deflection and thus inter-storey drift increased with structures' number of stories and depending on the earthquake PGA, some structures cases got destructed. Moreover, foundation rocking angle increased with structures' number of stories which was reflected by the increase in the amount of structural and distortion components in flexible-based structures compared to fixed-based structures.

5. References

[1] S. Kramer (1996), Geotechnical Earthquake Engineering, Prentice Hall, 1996.

[2] H. El Ganainy, and MH. El Naggar (2009), "Seismic performance of three-dimensional frame structures with underground stories", Soil Dynamics and Earthquake Engineering, Elsevier, vol. 29, no. 9, pp. 1249-1261, 2009. doi: 10.1016/j.soildyn.2009.02.003.

[3] A. Farghaly and H. Ahmed "Contribution of soil-structure interaction to seismic response of buildings", KSCE Journal of Civil Engineering, Springer, vol. 17, no. 5, pp. 959-971, 2013. doi: 10.1007/s12205-013-0261-9.

[4] B. Fatahi and SH. Tabatabaiefar, "Fully nonlinear versus equivalent linear computation method for seismic analysis of mid-rise buildings on soft soils", International Journal of Geomechanics, ASCE, vol. 14, no. 4, pp. 04014016:1-15, 2014. doi:.10.1061/(ASCE)GM.1943-5622.0000354.

[5] K. Galal and M. Naimi M, "Effect of conditions on the response of reinforced concrete tall structures to near fault earthquakes", The Structural Design of Tall and Special Buildings , John Wiley and Sons,vol. 17, no.3,2008. doi: 10.1002/tal.365.

[6] A. Hokmabadi, B. Fatahi and B. Samali, "Effect of soil-pile-structure interaction on seismic response of moment resisting buildings on soft soil", presented at the 3rd International Conference on New Development in Soil Mechanics and Geotechnical Engineering, Near East University, Nicosia, North Cyprus, 2012

[7] M. Iida, "Three-dimensional finite-element method for soil-building interaction based on an input wave field", International Journal of Geomechanics, ASCE, vol. 13, no. 4, pp. 430-440, 2012. doi:.10.1061/ (ASCE)GM.1943-5622.0000232.

[8] H. Krawinkler , R. Medina and B. Alavi, "Seismic drift and ductility demands and their dependence on ground motions", Engineering Structures, Elsevier., vo. 25, no. 5, pp. 637-653 ,2003. doi: 10.1016/S0141-0296(02)00174-8.

[9] Q. Nguyen, B. Fatahi and A. Hokmabadi, "The effects of foundation size on the seismic performance of buildings considering the soil-foundation-structure interaction", Structural Engineering and Mechanics, vol. 58, no. 6, pp. 1045-1075, 2016. doi: 10.12989/sem.2016.58.6.1045.

[10] Q. Nguyen, B. Fatahi and A. Hokmabadi, "Influence of size and load-bearing mechanism of piles on seismic performance of buildings considering soil-pile-structure interaction", International Journal of Geomechanics, ASCE, vol. 17, no. 7, pp. 04017007"1-22 , 2017. doi: 10.1061/(ASCE)GM.1943-5622.0000869.

[11] E. Safak (1995), "Detection and identification of soil-structure interaction in buildings from vibration recordings", Structural Engineering, ASCE, vo. 121, no. 5, pp. :889-906, 1995. doi:. 10.1061/(ASCE) 0733-9445(1995)121:5(899), 889–906.

[12] EA. Shehata, M. Ahmed and T. Alazrak, "Evaluation of soil-foundation-structure interaction effects on seismic response demands of multi-story mrf buildings on raft foundations", International Journal of Advanced Structural Engineering, Springer, 7(1):11-30, 2015. doi: 10.1007/s40091-014-0078-x.

[13] C. Spyrakos , P. Patel and F. Kokkinos, "Assessment of computational practices in dynamic soil-structure interaction", Computing in Civil Engineering, ASCE, vol. 3, no. 2, pp. 43-157, 1989. doi: 10.1061/(ASCE)0887-3801(1989)3:2.

[14] SH. Tabatabaiefar, B. Fatahi and B. Samali, "An empirical relationship to determine lateral seismic response of mid-rise building frames under influence of soil-structure interaction", The Structural Design of Tall and Special Buildings , vol. 23, no. 7, pp. 526-548, 2014. doi: 10.1002/tal.1058.

[15] SH. Tabatabaiefar, B. Fatahi and B. Samali, "Numerical and experimental investigations on seismic response of building frames under influence of soil-structure interaction", International Journal of Advanced Structural Engineering, Springer, vol. 17, no. 1, pp. 109-10, 2014. doi: 10.1260/1369-4332.17.1.109.

[16] A. Veletsos and A. Prasad, "Seismic interaction of structures and soils: stochastic approach", Structural Engineering, ASCE, vol. 115, no. 4, pp. 935-956, 1989. doi: 10.1061/(ASCE)07339445(1989)115:4.

[17] A. Veletsos and J. Meek, "Dynamic behaviour of building foundation system", Earthquake Engineering and Structural Dynamics, vol. 3, no. 2, pp. 121–138, 1974. doi: 10.1002/eqe.4290030203.

[18] G. Mylonakis and G. Gazetas, "Seismic soil-structure interaction: beneficial or detrimental", Earthquake Engineering, vol. 4, no. 3, pp. 277-301, 2000. doi: 10.1080/13632460009350372.

[19] Abaqus [Computer Software], Dassault Systèmes SIMULIA Corporation, Minneapolis, 2017.

[20] A. Chopra, Dynamics of structures theory and applications to earthquake engineering, Prentice Hall ,Inc. Fourth edition, 2011.

[21] HB. Seed and I. Idriss (1969), "Influence of soil conditions on ground motion during earthquakes", Soil Mechanics and Foundation Division, vol. 95, no. 2, pp. 99–137, 1969.

[22] M. Vucetic and R. Dobry, "Effect of soil plasticity on cyclic response", Geotechnical Engineering, vol. 117, no. 1, pp. 89, 1991. doi: 10.1061/(ASCE)0733-9410(1991)117:1(89).

[23] D. Park and Y. Hashash, "Soil damping formulation in nonlinear time domain site response analysis", Earthquake Engineering, vol. 8, no. 2, pp. 249–274, 2003.

[24] S. Ghosh S and EL. Wilson, Dynamic stress analysis of axi-symmetric structures under arbitrary loading, Report no. EERC 69-10, University of California, Berkeley, 1969.

[25] M. Rayhani and M. El Naggar, "Numerical modeling of seismic response of rigid foundation on soft soil", International Journal of Geomechanics, ASCE, vol. 8, no. 6, pp: 336-346, 2008. doi:10.1061/(ASCE)1532-3641(2008)8:6(336),336–346.

[26] S. Ismail, F. Kaddah and W. Raphael, "Seismic soil structure interaction of a midrise frame structure", presented at the 2nd GeoMEast International Congress and Exhibition on Sustainable Civil Infrastructures, November 2008. doi: 10.007/978-3-030-01923-5.

[27] BSSC, NEHRP Guidelines for the Seismic Rehabilitation of Buildings, 1997 Edition, Part 1: Provisions and Part 2: Commentary, Federal Emergency Management Agency, 1997.

[28] BSSC, NEHRP Recommended Seismic Provisions for New Buildings and Other Structures, Federal Emergency Management Agency, 2009.

[29] A. Hokmabadi, B. Fatahi and B. Samali, “Recording interstorey drifts of structures in time-history approach for seismic design of building frames”, Australian Journal of Structural Engineering, vol. 13, no. 2, 2012. doi: 10.7158/13287982.2012.11465107.

[30] MATLAB and Statistics Toolbox Release [Computer Software], The MathWorks, Inc., Natick, Massachusetts, United States, 2007.

[31] W. Shiming, G. Gang, “Dynamic soil-structure interaction for high-rise buildings”, Dynamic Soil-Structure Interaction Developemens in Geotechnical Engineering, El-Sevier, 1998.

[32] JE. Luco and HL. Wong, “Response of a rigid foundation to a spatially random ground motions”, Engineering Mechanics, vo. 14, no. 6, pp. 891-908, 1986. doi: 10.1002/eqe.4290140606.

[33] A. Veletsos, A. Prasad and WH. Wu, “Transfer functions for rigid rectangular foundations”, Earthquake Engineering Structural Dynamics, vo. 26, no. 1, pp. 5-17, 1997. doi: 10.1002/(SICI)1096-9845(199701)26:1<5::AID-EQE619>3.0.CO;2-X.

[34] M. Trifunac, SS. Ivanovic and M. Todorovska , “Apparent periods of a building. ii: time-frequency analysis”, Structural Engineering, ASCE, vo. 127, no. 5, pp. 527-357, 2001. doi: 10.1061/(ASCE)0733-9445(2001)127:5(527).

[35] M. Trifunace, M. Todorovska and T. Hao, “Full-scale experimental studies of soil-structure interaction - a review”, presented at the 2nd U.S.-Japan Workshop on Soil-Structure Interaction, Tsukuba City, Japan, March 2001.

IOP Conf. Series: Materials Science and Engineering **809** (2020) 012012 doi:10.1088/1757-899X/809/1/012012

Effect of Raft and Column Sizes on the Seismic Soil Structure Interaction Performance of Fifteen Storey Midrise Frame Structures

Sahar A. Ismail[a], Fouad K. Kaddah[b] and Wassim E. Raphael[c]

Université Saint-Joseph de Beyrouth / Department of Civil & Environmental Engineering, Beyrouth, Lebanon

[a] sahar.ismail@net.usj.edu.lb , [b] fouad.kaddah@usj.edu.lb, [c]wassim.raphael@usj.edu.lb

Abstract. Seismic design of structures is usually executed by performing 1D free-field soil response analysis using design spectral acceleration charts given by codes. This method ignores soil structure interaction effects thus affecting structure-foundation-soil system behavior leading to imprecise design. Soil structure interaction refers to the dynamic behavior between the structure, underlying foundation and soil block. To understand the behavior of a 15 storey midrise concrete seismic frame structure rested on raft foundation and founded on silty sandy soil block, 3D nonlinear time history finite element simulations using Abaqus were performed. The effects of raft and column sizes were considered by modeling flexible and fixed based structures. All cases were hit at the bottom by El-Centro (1940) and Northridge (1994) earthquakes. The results were presented in terms of storey lateral deflection, inter-storey drift, shear force, foundation rocking and response spectrum. The results showed that the combination between the structure-foundation-soil natural frequencies along with the soil type used dictate the behavior of the midrise structure as well as SSI effects under the different raft and column sizes used. Therefore, a detailed study considering all above parameters must be performed by engineers to ensure safe and cost-effective designs.

Notation List: c= soil apparent cohesion value[kPa], γ_d=dry unit weight[kN/m^3], ρ =density[kg/m^3], Φ= internal friction angle[°],Ψ=dilation angle[°], E=Young Modulus[MPa], G= shear modulus[MPa], f'_c=concrete compressive strength[MPa] , e=void ratio, ν=poisson's ratio, PI=plasticity index[%], V_s=shear wave velocity[m/s], Υ_c= cyclic shear strain[%], A=acceleration[g], PGA=peak ground acceleration[g], S_a=spectral acceleration[g], ζ=damping ratio[%],d_i= deflections at i level[mm], h=the story height[m], α=mass damping factor, β=stiffness damping factor.

1. Introduction

Scientists and engineers use "seismic hazard" and "seismic risk" terms to describe earthquakes and their effects on structures, soils and lives. "Seismic hazard" describes the natural phenomena left by an earthquake such as ground shaking, fault rupture and structural hazards though "seismic risk" describes the damage to lives and structures due to a seismic hazard. Thus, depending on the vulnerability of a structure to mitigate the effects of an earthquake, a seismic hazard can either increase or decrease a seismic risk. Earthquake geo-hazard to ground shaking can be expressed in terms of ground response, site effects, settlements, etc. Noting that although seismic waves travel in rock over their most trips to the earth's surface, the final portion of their trip is very important. The seismic waves finish their trip by propagating through soil deposits. Therefore, depending on the characteristics of these soil deposits, the waves can be either get amplified (mainly in cohesive soils) or attenuated (mainly in cohesionless soils) [1]. Thus, the seismic response of a structure is affected by the dynamic response of the soil medium where the structure and the foundation are founded called soil structure interaction (SSI). Codes recommend designing structures base on 1D spectral acceleration curves neglecting SSI effects as a mean to improve the factor of safety. However, neglecting SSI in structures' design lead to a decrease in structure's fundamental period, damping ratio and lateral displacement with an increase or a decrease in structure's base shear depending on the soil spectral acceleration.

The effects of earthquakes on soils and structures have been studied by several researchers since early 1960s. Nevertheless, it wasn't till after 1971 M_w=6.6 San Fernando earthquake that building codes started considering local site effects and research in this field expanded using analytical, experimental and numerical solutions [2, 3, 4, 5, 6, 7, 8, 9, 10, 11, 12, 13, 14, 15, 16, 17, 18, etc]. Since then, several significant earthquakes occurred around the world and expanded researchers' knowledge in this area such as 1994 M_w=6.7 Northridge and 1995 M_w=7.5 Kobe earthquakes. These earthquakes left thousands of people dead and destructed hundred thousand of buildings leaving thousands of people homeless.

The finite element software SAP, Etabs, Ansys, Abaqus, etc and finite difference software like Flac are being used to model SSI problem using the direct method of analysis. Ref. [6], Ref. [9], Ref. [12] and Ref. [15] are among many others who showed that due to SSI effects, lateral displacement and thus inter-storey drifts of flexible-structures are greater than those of fixed-based structures. The effect of foundation type was studied by Ref. [16], Ref. [17], Ref. [9] and Ref. [10], etc. Ref. [16] and Ref. [17] found that earthquake type (near or far-field) affects soil amplification within the same soil type. Moreover, they showed that earthquake motion characteristics are affected by the structure as well as by the foundation type. Nevertheless, SSI effects can be neglected in designs when soils have shear wave velocities below 600m/s [5, 19] and in cohesionless not cohesive soils [14, 20]. In addition, Ref. [9] demonstrated that as the raft foundation size decreased, the inertial forces in the structure decreased thus, the spectral acceleration reduced significantly due to the increase in structure natural frequency leading to an increase in lateral deflection and foundation rocking.

All above researches were conducted on soft soils. This paper deals with the dynamic behavior of 15 storey concrete frame structure founded on raft foundation and silty sandy soil and analyses the effects of raft and column sizes including soil structure interaction effects. Fully 3D nonlinear seismic finite element simulations under the influence of two strong ground motions: El-Centro (1940) and Northridge (1994) were conducted using Abaqus. The results were presented in terms of storey lateral displacement, inter-storey drifts, shear force, foundation rocking and response spectrum and were used to compare the behavior of flexible to fixed-based structures. This paper highlights the importance of considering soil structure interaction effects and the influence of raft and column sizes on structures' design in terms of safety as well as cost-effectiveness.

2. Finite element model

To understand the behavior of 15 storey midrise concrete seismic frame structures rested on raft foundation and silty sandy soil, the effects of raft and column sizes were investigated by performing three-dimensional nonlinear finite element simulations using the commercial software Abaqus [21]. Flexible (S-N-C models with N referring to the raft size used and C to the column size used in each case) and fixed base structures (S-C models) were modeled by the direct method of analysis where the entire structure-raft-soil system was simulated at the same time as shown in Figure 1. Moreover, each modeled case was hit at the bottom by El-Centro (1940) M_w=6.9, PGA=0.318g near field and Northridge (1994) M_w=6.7, PGA=0.843g far field earthquakes as presented in Figure 2. The modeled configuration details and material properties used in this paper are presented in Table 1 and 2. The structure's beams and columns were modeled using two-node linear beam B31 with 9400 elements while the structure's slabs were modeled using four-node double curved shell S4 with 100 elements. The raft and soil block were modeled using eight-node linear brick, reduced integration, hourglass control continuum solid elements C3D8R: the raft with around 4800 elements and the soil block with 96500 elements. Moreover, uniformly distributed live and dead load of 2.5 kN/m^2 was applied to the structures' floors. The bottom soil boundary was assumed fixed while the horizontal soil boundaries were modeled as quiet boundary

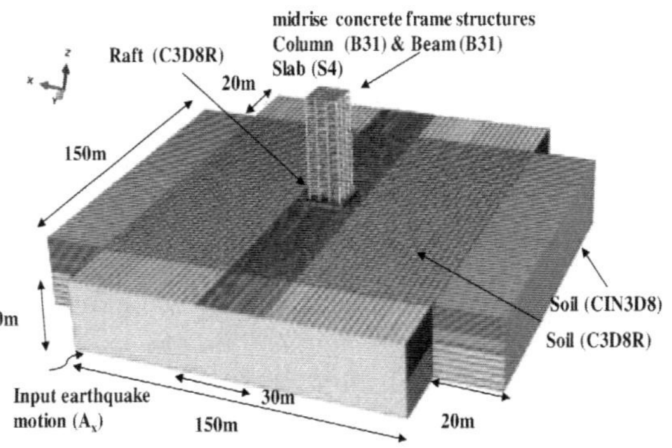

Figure 1. Geometry and model mesh distribution

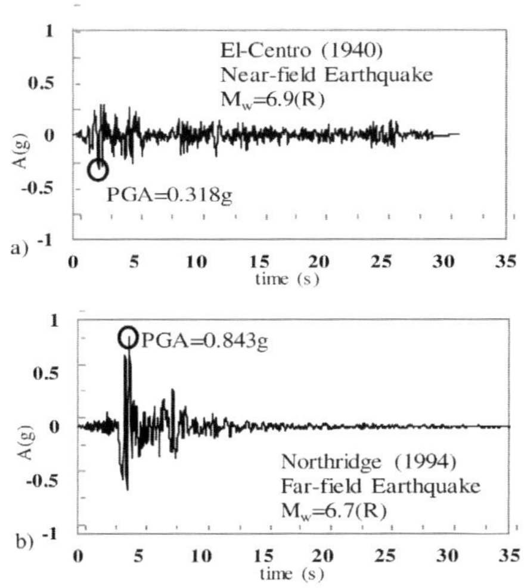

Figure 2. Acceleration with respect to time of a) El-Centro (1940) and b) Northridge (1994) earthquake

Table 1. Model configuration

Number of Bays	Storey Height [m]	Bay Width [m]	Column [cm]	Beam [cm]	Slab [m]
3	3	5	50X50, 50X100	50X50	5X5X0.25
Soil Dimensions			Raft Dimension		
Height [m]	Length [m]	Width [m]	Length [m]	Width [m]	Thickness [m]
30	150	150	20(1.3B), 22.5(1.5B), 30(2B)	20(1.3B), 22.5(1.5B),30(2B)	1.5

conditions using one-way infinite brick elements CIN3D8 to simulate the absorbed energy from the unbounded soil domain. Element interactions were accounted for by tying structures' beams and slabs as well as columns and raft and raft and soil surfaces whereas embedded columns were placed in flexible models' rafts. Noting that each modeled case took around 60 hours to be completed using fast computational facilities at Université Saint-Joseph de Beyrouth.

Structural elements were modeled using elastic-perfectly plastic material while considering 5% Rayleigh material damping. Then, the mass damping factor (α) and the stiffness damping factor (β) were calculated based on first and second mode structural frequencies (f_i and f_j in rad/s) obtained using linear perturbation procedure and Lancoz method available in Abaqus (Table 3):

$$\zeta = \frac{1}{2f_n}\alpha + \frac{f_n}{2}\beta \quad (1), \qquad \alpha = 2\zeta_i \frac{f_i f_j}{f_i + f_j} \quad (2) \text{ and} \qquad \beta = 2\zeta_i \frac{1}{f_i + f_j} \quad (3)$$

Moreover, dense silty sandy soil having a shear wave velocity of 63 m/s was used in this paper. To calculate its shear modulus (G) and damping ratio (ζ), equivalent linear method was used. In this method, the maximum cyclic shear strain for each soil element was determined by performing linear analysis of preliminary assumed values of G and ζ [22,23,24]. Then, using backbone curves given by Ref. [25] that relates G and ζ to the cyclic shear strain for different plasticity indices for cohesive soils, new values of G and ζ were obtained. These values were therefore applied in the numerical model and this phase was repeated until no further change in G and ζ were obtained. After that, using linear perturbation procedure and Lancoz method available in Abaqus, first and second mode soil frequencies were determined to calculate soil Rayleigh damping coefficients (Table 3). Noting that each earthquake is unique, thus each earthquake possesses its own damping coefficients.

Table 2. Material Properties

Concrete	E[GPa]	ν	ρ [kg/m^3]	f'$_c$ [MPa]
	14	0.2	2400	20
	Silty Sandy Soil			
	γ_{max}[%]*	G/G$_{max}$*	ζ[%]*	α & β*
El-Centro (1940)	1.72	0.085	20.5	0.63, 0.066
Northridge (1994)	1.82	0.0769	21	0.65, 0.068
E [MPa]	ν	γ_d[kN/m^3]	e	* after [22]
20	0.3	18.98	0.4	
φ°	ψ°	c [kPa]	V$_s$[m/s^2]	PI [%]
35	10	50	63	15

Table 3. Structures' natural frequencies

Reference name	f1[Hz]	f2[Hz]	α*	β*
S15-C-0.5-0.5m	0.47	1.45	0.22	0.008
S15-C-0.5-1m	0.54	1.64	0.25	0.007
* after [25]				

3. Results and analysis

To study the effects of raft and column sizes of a 15 storey seismic concrete frame structure, the size of the raft foundation was varied between 1.3, 1.5 and 2B with B=15m being the width of the structure under 2 different column sizes: 50 X 50 cm and 50 X 100 cm with 100 cm being in the direction of the earthquake load. All raft foundations had thicknesses of 1.5m and the soil boundary limits were chosen after a series of parametric simulations and the recommendations of Ref. [26], Ref. [27] and Ref. [28]. The results, presented in Figures 3 to 5 and Tables 4 in terms of storey lateral deflection, inter-storey drifts, shear force, foundation rocking and response spectrum, compare the results of flexible to fixed base cases. Noting that all cases were hit at the bottom by El-Centro (1940) and Northridge (1994) earthquakes. In addition, storey lateral deflection plots, plotted when the maximum lateral deflection occurs at top of the structure regardless what time it occurred [28, 29], were measured relative to the base of the structures.

3.1 Lateral deflection and Inter-storey drifts

The results presented in Figures 3 and 4 show that flexible structures present higher lateral deflections than fixed base structures showing SSI effects which is accordance with Ref. [6], Ref. [9], Ref. [12] and Ref. [15]. Moreover, 1.5 and 2B raft sizes present similar deflection curves under both earthquakes. Also, increasing the raft size from 1.3 to 2B for C-0.5-0.5m cases increased top storey lateral deflection by a ratio of 1.41 and 1.12 under El-Centro (1940) and Northridge (1994) earthquakes respectively. However, increasing the raft size from 1.3 to 2B for C-0.5-1m cases decreased top storey lateral deflection by a ratio of 0.56 and 0.86 under El-Centro (1940) and Northridge (1994) earthquakes respectively.

Therefore, lateral deflection increased for 1.3B cases while it decreased for 2B cases when increasing the column sizes under both excitations. For example, lateral deflection at top of the structure decreased by a ratio of 0.69 from S15 C-0.5-0.5m to C-0.5-1m while it increased by a ratio of 1.24 from S15-1.3B C-0.5-0.5m to C-0.5-1m and then decreased by a ratio of 0.49 from S15-2B C-0.5-0.5m to C-0.5-1m.

These lateral deflections are used to calculate the level of damage of a structure from a seismic excitation. This damage parameter called inter-storey drift is calculated as the difference between the lateral deflection of two stories divided by the storey height:

$$drift = (d_i - d_{i+1})/h \quad (4)$$

It is divided into 5 categories: if inter-storey drift < 0.2%: fully operational, < 0.5%: operational, < 1.5%: life safe, < 2.5%: near collapse and > 2.5%: collapse with life safe category being the acceptable limit by most

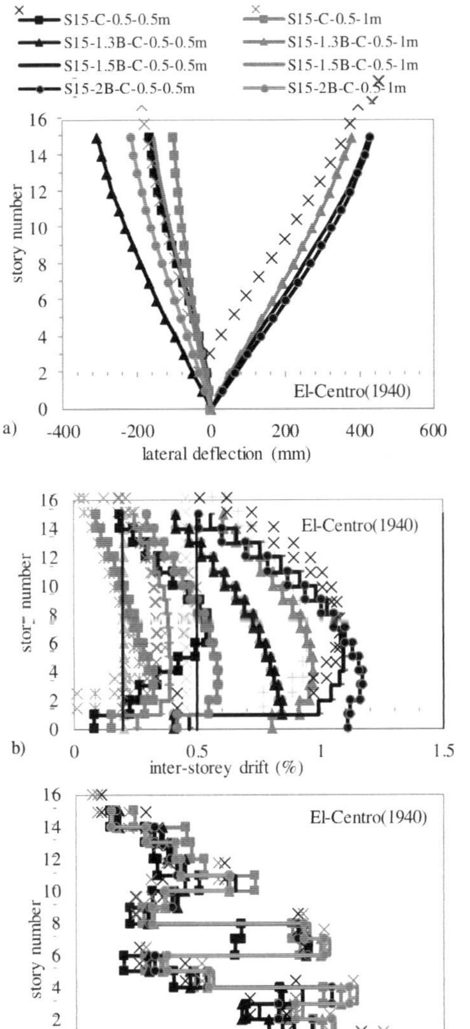

Figure 3. a) Lateral deflection, b) inter-storey drift and c) maximum shear force distribution of flexible and fixed-base models under El-Centro (1940) earthquake

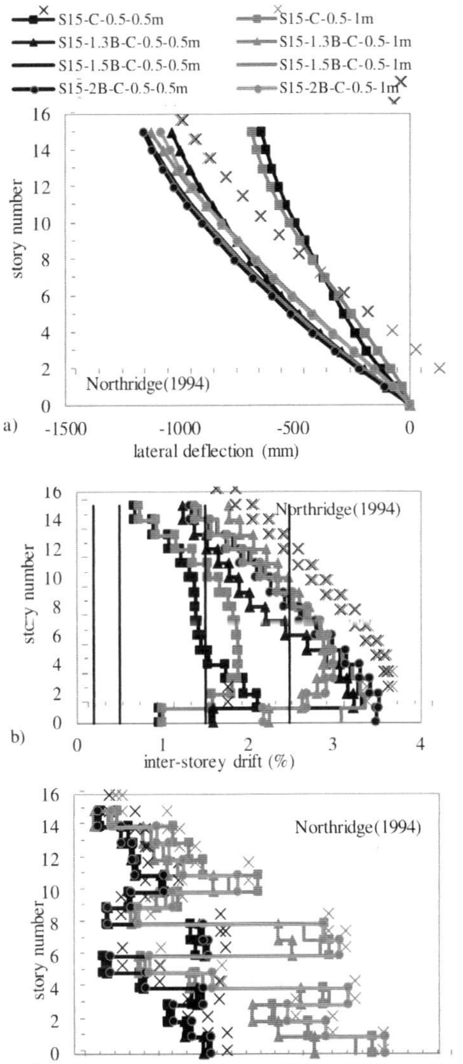

Figure 4. a) Lateral deflection, b) inter-storey drift and c) maximum shear force distribution of flexible and fixed-base models under Northridge (1994) earthquake

seismic codes [30, 31]. As shown in Figure 3 and 4, increasing raft size pushed inter-storey drift curves into higher categories where it reached collapse under Northridge (1994) earthquake. Therefore, even though both earthquakes had similar magnitudes, the increase in raft size was more pronounced under El-Centro (1940) earthquake while the safety of the structure was more affected under Northridge (1940) earthquake having higher PGA value.

Soil structure interactions are divided into kinematic and inertial interactions. Kinematic interactions are caused by the stiffness of the structure while inertial interactions are caused by the mass of the structure and the foundation. Thus, increasing the size of the raft or the column increases the mass of the structure-foundation

IOP Conf. Series: Materials Science and Engineering **809** (2020) 012012 doi:10.1088/1757-899X/809/1/012012

system ending in extra inertial interaction manifested by the increase in the system's base motion. However, this motion is related to the ability of the system to match kinematic interaction [1] and depending on the soil stiffness, kinematic interaction can cause a decrease in structure's base motion [18]. In general, increasing raft size tends to decrease structure's lateral deflection. Nevertheless, due to the use of silty sandy soil, the combination between the structure-foundation-soil different natural frequencies as well as wave attenuation, earthquake PGA and column stiffness caused larger foundations to attract more energy than smaller foundations under C-0.5-0.5m cases. In other words, SSI effects were more pronounced at larger foundation structures under C-0.5-0.5m cases and the opposite under C-0.5-1m cases. The results presented in this paper are not entirely in parallel with Ref [9] who due to his use of clayey soil, seismic wave was amplified when it reached top of the soil profile and led to the decrease in lateral deflection and inter-storey drift values.

3.2 Foundation Rocking

Lateral deflections and inter-storey drifts values depend on the amount of structural rocking component experienced by the structure. They are caused by the inertial forces formed in the structure that cause compression on the first side of the foundation along with a possible uplift on the other side. As shown in Table 4, the increase in raft size caused a decrease in foundation rocking under both column sizes like the results obtained by [9]. The foundation rocking angle decreased from 0.22° to 0.08° when increasing the raft size from 1.3 to 2B for C-0.5-0.5m under El-Centro (1940) ground motion. Nevertheless, using Ref [32] and Ref [33] relationship that relates the foundation rocking angle to the amount of structure's distortion and rocking components, the results showed that the increase in raft size caused a decrease in the amount of distortion component with an increase in the amount of rocking component. For example, taking C-0.5-1m under Northridge (1994) excitation case, the maximum lateral displacement of S15 was equal to 687.8mm corresponding to the distortion component. This lateral displacement increased in S15-1.3B to 1125.38mm with a rocking angle of 0.78° thus dividing it into 877.80mm as distortion component (78%) and 247.58mm (22%) as rocking component. Then, it decreased in S15-2B to 1084.31mm with a rocking angle of 0.27° thus dividing this latter into 292.76mm as distortion component (27%) and 791.55mm (73%) as rocking component.

On the other hand, the results showed that the increase in column stiffness (its size) caused an increase in structure's distortion component and a decrease in its rocking component. Taking S15-2B Northridge (1994) earthquake case as an example, the maximum lateral displacement of S15 C-0.5-0.5m was equal to 643.56mm (distortion component), it increased in S15-2B to 1159.6mm: 197.13mm (17% distortion component) and 962.5mm (83% rocking component). While the maximum lateral displacement of S15 C-0.5-1m was equal to 687.8mm (distortion component), it increased in S15-2B to 1084.31mm: 292.76 (27% distortion component) and 791.55mm (73% rocking component). Noting that for the same column-raft size case, foundation rocking angle was greater under Northridge (1994) than under El-Centro (1940) earthquake since Northridge (1994) possesses a higher PGA causing a larger earthquake wave beneath the system. Thus, the relative structural displacement, divided into structural and rocking components, is related to SSI effects expressed by the interaction between the column stiffness, raft size, soil type and earthquake wave.

Table 4. Maximum foundation rocking angle [°]

	Reference name	El-Centro (1940)	Northridge (1994)
	S15-1.3B	0.22	0.44
C-0.5X0.5m	S15-1.5B	0.15	0.31
	S15-2B	0.08	0.17
	S15-1.3B	0.23	0.78
C-1X0.5m	S15-1.5B	0.13	0.58
	S15-2B	0.07	0.27

3.3 Level shear force and response spectrum

Shear force and response spectrum plots were plotted in Figures 3 to 5 noting that to calculate the maximum absolute shear force at every storey, all shear forces produced by all columns in each storey were summed up. The results, presented in Table 5, showed that raft size slightly affected average level shear force values under both earthquakes for a given column size. However, the increase in column size decreased average level shear force value of flexible to fixed base models for a given raft size. For example, the average level shear force of flexible to fixed base models increased from 0.99 to 1.03 and from 1.20 to 1.28 from 1.3B to 2B for C-0.5-0.5m cases under El-Centro (1940) and Northridge (1994) earthquakes respectively as detailed in Table 5. The increase in column size increased structure's natural frequency (Table 1) resulting in more SSI effects pronounced by lower average level shear force values [34].

Using a Fortran code, the variation of pseudo-acceleration response spectrum plots (S_a) with different natural periods at 5% damping of horizontal accelerations formed at structures' base, at top of the soil profile (Free Field: FF) and of the input earthquake were created (Figure 5). Also, using MATLAB [35], Fourier transformation of the strong ground motions produced at top of the soil surface (FF) and at the base of the structure-foundation system were generated also at 5% damping (not shown due to page size limits). The plots showed that due to the use of silty sandy soil, which is a combination of cohesive and cohesionless soil, input earthquake wave was attenuated when it reached top of the soil profile (FF). Noting that a small wave amplification happened between FF and structures' base illustrating SSI effects which is in parallel with [27]. Also, increasing raft and column size slightly affected response spectrum curves like the results obtained by [9] who obtained a negligible response spectrum variation when increasing the raft size from 1.1 to 2B.

Table 5. Average level shear force of flexible to fixed base models

Column size	C-0.5X0.5m			C-0.5X1m		
Model S15-	1.3B	1.5B	2B	1.3B	1.5B	2B
El-Centro (1940)	1.20	1.27	1.28	0.96	0.96	0.96
Northridge (1994)	0.99	1.01	1.03	0.80	0.92	0.98

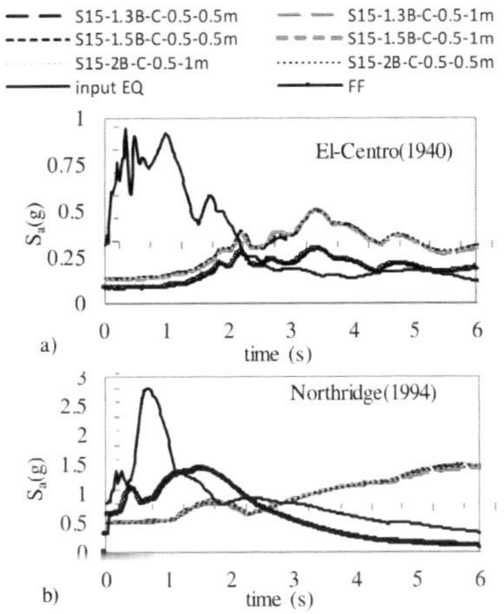

Figure 5. Acceleration response spectrum with 5% damping ratio for the midrise frame structure under the influence of a) El-Centro (1940) and b) Northridge (1994) earthquakes

4. Conclusions

The behavior of 15 storey seismic midrise concrete frame structure rested on raft foundation and founded on silty sandy soil under the effects of raft and column sizes while considering SSI effects was investigated in this paper. The results showed that the size of raft as well as column stiffness can affect the performance of the structure. Larger foundations can attract more energy than smaller foundations. However, the increase in raft size and the decrease in column size cause a decrease in the amount of distortion component with an increase in the amount of rocking component. Nevertheless, raft size slightly affects shear force values and response spectrum curves while the increase in column size decreases average level shear force values. These results were related to the use of silty sandy soil as well as the relation between the structure-foundation-soil different natural frequencies, earthquake wave attenuation and PGA along with column stiffness and raft size. Thus, engineers should carefully consider all above parameters in design to ensure safety and cost effectiveness.

5. References

[1] S. Kramer (1996), Geotechnical Earthquake Engineering, Prentice Hall, 1996.

[2] H. El Ganainy, and MH. El Naggar (2009), "Seismic performance of three-dimensional frame structures with underground stories", Soil Dynamics and Earthquake Engineering, Elsevier, vol. 29, no. 9, pp. 1249-1261, 2009. doi: 10.1016/j.soildyn.2009.02.003.

[3] A. Farghaly and H. Ahmed "Contribution of soil-structure interaction to seismic response of buildings", KSCE Journal of Civil Engineering, Springer, vol. 17, no. 5, pp. 959-971, 2013. doi: 10.1007/s12205-013-0261-9.

[4] B. Fatahi and SH. Tabatabaiefar, "Fully nonlinear versus equivalent linear computation method for seismic analysis of mid-rise buildings on soft soils", International Journal of Geomechanics, ASCE, vol. 14, no. 4, pp. 04014016:1-15, 2014. doi:.10.1061/(ASCE)GM.1943-5622.0000354.

[5] K. Galal and M. Naimi M, "Effect of conditions on the response of reinforced concrete tall structures to near fault earthquakes", The Structural Design of Tall and Special Buildings , John Wiley and Sons,vol. 17, no.3,2008. doi: 10.1002/tal.365.

[6] A. Hokmabadi, B. Fatahi and B. Samali, "Effect of soil-pile-structure interaction on seismic response of moment resisting buildings on soft soil", presented at the 3rd International Conference on New Development in Soil Mechanics and Geotechnical Engineering, Near East University, Nicosia, North Cyprus, 2012

[7] M. Iida, "Three-dimensional finite-element method for soil-building interaction based on an input wave field", International Journal of Geomechanics, ASCE, vol. 13, no. 4, pp. 430-440, 2012. doi:.10.1061/(ASCE)GM.1943-5622.0000232.

[8] H. Krawinkler , R. Medina and B. Alavi, "Seismic drift and ductility demands and their dependence on ground motions", Engineering Structures, Elsevier., vo. 25, no. 5, pp. 637-653 ,2003. doi: 10.1016/S0141-0296(02)00174-8.

[9] Q. Nguyen, B. Fatahi and A. Hokmabadi, "The effects of foundation size on the seismic performance of buildings considering the soil-foundation-structure interaction", Structural Engineering and Mechanics, vol. 58, no. 6, pp. 1045-1075, 2016. doi: 10.12989/sem.2016.58.6.1045.

[10] Q. Nguyen, B. Fatahi and A. Hokmabadi, "Influence of size and load-bearing mechanism of piles on seismic performance of buildings considering soil-pile-structure interaction", International Journal of Geomechanics, ASCE, vol. 17, no. 7, pp. 04017007"1-22 , 2017. doi: 10.1061/(ASCE)GM.1943-5622.0000869.

[11] E. Safak (1995), "Detection and identification of soil-structure interaction in buildings from vibration recordings", Structural Engineering, ASCE, vo. 121, no. 5, pp. :889-906, 1995. doi:. 10.1061/(ASCE) 0733-9445(1995)121:5(899), 889–906.

[12] EA. Shehata, M. Ahmed and T. Alazrak, "Evaluation of soil-foundation-structure interaction effects on seismic response demands of multi-storey mrf buildings on raft foundations", International Journal of Advanced Structural Engineering, Springer, 7(1):11-30, 2015. doi: 10.1007/s40091-014-0078-x.

[13] C. Spyrakos , P. Patel and F. Kokkinos, "Assessment of computational practices in dynamic soil-structure interaction", Computing in Civil Engineering, ASCE, vol. 3, no. 2, pp. 43-157, 1989. doi: 10.1061/(ASCE)0887-3801(1989)3:2.

[14] SH. Tabatabaiefar, B. Fatahi and B. Samali, "An empirical relationship to determine lateral seismic response of mid-rise building frames under influence of soil-structure interaction", The Structural Design of Tall and Special Buildings , vol. 23, no. 7, pp. 526-548, 2014. doi: 10.1002/tal.1058.

[15] SH. Tabatabaiefar, B. Fatahi and B. Samali, "Numerical and experimental investigations on seismic response of building frames under influence of soil-structure interaction", International Journal of Advanced Structural Engineering, Springer, vol. 17, no. 1, pp. 109-10, 2014. doi: 10.1260/1369-4332.17.1.109.

[16] AS. Hokmabadi, B. Fatahi, "Influence of foundation type on seismic performance of buildings considering soil–structure interaction", International Journal of Structural Stability and Dynamics, vo. 16, no. 8, 2016. doi: 10.1142/S0219455415500431

[17] AS. Hokmabadi, B. Fatahi, B. Samali, "Recording inter-storey drifts of structures in time-historey approach for seismic design of building frames", Australian Journal of Structural Engineering, vo. 13, pp:175-179.

[18] A. Veletsos and A. Prasad, "Seismic interaction of structures and soils: stochastic approach", Structural Engineering, ASCE, vol. 115, no. 4, pp. 935-956, 1989. doi: 10.1061/(ASCE)07339445(1989)115:4.

[19] A. Veletsos and J. Meek, "Dynamic behaviour of building foundation system", Earthquake Engineering and Structural Dynamics, vol. 3, no. 2, pp. 121–138, 1974. doi: 10.1002/eqe.4290030203.

[20] G. Mylonakis and G. Gazetas, "Seismic soil-structure interaction: beneficial or detrimental", Earthquake Engineering, vol. 4, no. 3, pp. 277-301, 2000. doi: 10.1080/13632460009350372.

[21] Abaqus [Computer Software], Dassault Systèmes SIMULIA Corporation, Minneapolis, 2017.

[22] HB. Seed and I. Idriss (1969), "Influence of soil conditions on ground motion during earthquakes", Soil Mechanics and Foundation Division, vol. 95, no. 2, pp. 99–137, 1969.

[23] M. Vucetic and R. Dobry, "Effect of soil plasticity on cyclic response", Geotechnical Engineering, vol. 117, no. 1, pp. 89, 1991. doi: 10.1061/(ASCE)0733-9410(1991)117:1(89).

[24] D. Park and Y. Hashash, "Soil damping formulation in nonlinear time domain site response analysis", Earthquake Engineering, vol. 8, no. 2, pp. 249–274, 2003.

[25] A. Chopra, Dynamics of structures theory and applications to earthquake engineering, Prentice Hall ,Inc. Fourth edition, 2011.

[26] S. Ghosh S and EL. Wilson, Dynamic stress analysis of axi-symmetric structures under arbitrary loading, Report no. EERC 69-10, University of California, Berkeley, 1969.

[27] M. Rayhani and M. El Naggar, "Numerical modeling of seismic response of rigid foundation on soft soil", International Journal of Geomechanics, ASCE, vol. 8, no. 6, pp: 336-346, 2008. doi:10.1061/(ASCE)1532-3641(2008)8:6(336),336–346.

[28] S. Ismail, F. Kaddah and W. Raphael, "Seismic soil structure interaction of a midrise frame structure", presented at the 2nd GeoMEast International Congress and Exhibition on Sustainable Civil Infrastructures, November 2008. doi: 10.007/978-3-030-01923-5.

[29] A. Hokmabadi, B. Fatahi and B. Samali, "Recording interstorey drifts of structures in time-historey approach for seismic design of building frames", Australian Journal of Structural Engineering, vol. 13, no. 2, 2012. doi: 10.7158/13287982.2012.11465107.

[30] BSSC, NEHRP Guidelines for the Seismic Rehabilitation of Buildings, 1997 Edition, Part 1: Provisions and Part 2: Commentary, Federal Emergency Management Agency, 1997.

[31] BSSC, NEHRP Recommended Seismic Provisions for New Buildings and Other Structures, Federal Emergency Management Agency, 2009.

[32] M. Trifunac, SS. Ivanovic and M. Todorovska , "Apparent periods of a building. ii: time-frequency analysis", Structural Engineering, ASCE, vo. 127, no. 5, pp. 527-357, 2001. doi: 10.1061/(ASCE)0733-9445(2001)127:5(527).

[33] M. Trifunace, M. Todorovska and T. Hao, "Full-scale experimental studies of soil-structure interaction - a review", presented at the 2nd U.S.-Japan Workshop on Soil-Structure Interaction, Tsukuba City, Japan, March 2001

[34] W. Shiming, G. Gang, "Dynamic soil-structure interaction for high-rise buildings", Dynamic Soil-Structure Interaction Developemens in Geotechnical Engineering, El-Sevier, 1998.

[35] MATLAB and Statistics Toolbox Release [Computer Software], The MathWorks, Inc., Natick, Massachusetts, United States, 2007.

IOP Conf. Series: Materials Science and Engineering **809** (2020) 012013 doi:10.1088/1757-899X/809/1/012013

Study on Working Behavior of Cage-type Flexible Retaining Wall

Liu Jie[1, a], Chen Yanyu[2, b], Geng Yehan[3, c], Li Kaixuan[4, d]

China Institute of Water Resources and Hydropower Research, Beijing

[a] lj_iwhr@163.com, [b] imexb@126.com, [c] gyh916@126.com, [d] jingnianlkx@163.com

Abstract. Cage-type flexible retaining wall is a new type of anti-seepage structure with broad market prospects due to its low cost and convenient installation. In order to study the working behavior of flexible cage-type retaining wall in flood control during the flood season, a three-dimensional finite element model of the structure was established, and the stress and deformation characteristics of the structure were analyzed. The results show that the maximum tensile stress is 17.1MPa, the maximum compressive stress is 12.6MPa, and the maximum structural deformation is 0.35mm, which are less than the allowable value, meaning the stress and deformation of the structure can meet the requirements of engineering construction. The reliability of the structure is verified by the full-scale model water retaining test. The vertical flow frame on the backwater side is a relatively weak part of the structure. When designing the structure, the cross-sectional dimensions of the vertical flow frame on the upstream side and the downstream flow frame can be reduced appropriately, while the cross-sectional dimensions of the vertical flow frame on the backwater side can be increased appropriately, so as to give better play to the material characteristics and save project cost.

1. Introduction

Removable flood wall has been widely used in flood control projects such as river dikes, port terminals, railway tunnel portals, highway culvert portals, civil air defense portals, etc. due to the advantages of fast installation and disassembly, easy stacking and collection, strong water pressure resistance and wide application range, etc.[1] According to the different rigidity of the flood wall, the removable flood wall can be divided into rigid and flexible ones. Rigid removable flood walls are widely used abroad. For example, removable aluminum alloy removable flood walls have been installed in the Czech Prague riverbank, Koln of Germany and Hungary, etc. Meanwhile, removable rigid flood walls are also used in China, such as the removable flood wall built above the designed water level of Wuhan Han River estuary shoreline[2], However, the high cost of rigid removable flood wall has limited its scope of use.

Compared with the rigid retaining wall, the new type of cage-type flexible retaining wall not only has the characteristics of fast installation, removability and small storage space as that of removable retaining wall, but also has a filling cage-type structure (an open cage structure consisting of square steel and wire mesh) which can use geomembrane to prevent permeation in the interior and on the upstream face. Geomembrane has been used in domestic water conservancy projects ever since 1980s and 1990s[3,4]. After 30 years of development, it has become a kind of low-cost and high-quality anti-seepage material widely used in the hydraulic seepage projects[5,6,7]. A built-in water bag made of geomembrane is laid in the cage-type flexible retaining wall, and water, soil or sediment is filled according to the site conditions to enhance the structural stability. Its cost is 30% lower than that of the rigid removable retaining wall. During the installation, fixed foundations are not required to be pre-casted on the shore, which further reduces the cost.

Many scholars have carried out relevant researches on the working behavior of removable rigid flood wall. Duan Yajuan[8] performed the water retaining test of rigid flood wall. The test results show that the midspan of baffle has the largest deflection and it is the weak part of the structure, so it is necessary to strengthen the reinforcement. Zhaxi Zhuoma[2] studied the failure mechanism and deformation characteristics of the baffle and columns of the IBS removable flood wall and pointed out that ENAW-6063T6 aluminum alloy material should be used, and the maximum span of its structure should not exceed 3m. Guo Lei et al.[9] studied the leakage characteristics of removable flood wall, which shows that the structure will leak when the water head exceeds 0.8m; the leakage will accelerate when the water head exceeds 1.70m. And during the actual application, the leakage of wall body and the installation accuracy of embedded parts are not high, which will aggravate the leakage of the system. Wu Junjun et al.[1] studied the working behavior of removable flood wall through field flood control water filling test and numerical simulation, and pointed out that the rigid flood wall of light aluminum alloy column could ensure the safety of flood control when the water filling height was not more than 3m. Zhao Gengrun et al.[10] calculated the current situation and long-term overflowing risk of flood wall in Shanghai Section of Huangpu River by using the Monte Carlo method on account of the overtopping risk of the flood wall in Shanghai Section of Huangpu River. The results show that the flood wall during the flood season has an insufficient elevation, which brings great hidden danger to flood control safety in urban area. At present, the research on the working behavior of cage-type flexible retaining wall has not been reported in the literature. Cage-type flexible retaining wall has broad application prospects due to its convenient construction, low cost, safety and reliability, and systematic research on its working behavior is required.

In terms of the cage-type flexible retaining wall, a new type of flood control structure, this paper uses ABAQUS finite element analysis software to establish a three-dimensional solid model, and studies and optimizes the stress and deformation law of the water retaining wall under different water retaining conditions, so as to provide theoretical support for the application of the cage-type flexible retaining wall.

2. Structural Style

The main structure of cage-type flexible retaining wall is composed of steel frame and steel wire mesh. The frame structure element can be made up of $1m \times 1m \times 1m$ or $1.5m \times 1.5m \times 1.5m$ square elements, and the frame element can be spliced along the horizontal, longitudinal and vertical directions. According to the needs of different water retaining heights, it can be spliced into different structural styles. Fixed connections are made between different elements by bolt pins, allowing a certain degree of deformation. The whole structure shows flexible characteristics and strong resistance to deformation.

Figure. 1 is the structural style and calculation model of the cage-type flexible retaining wall. Figure. 1.a shows a cross frame, consisting of steel frame, steel wire mesh and connecting framework. The square steel frame is welded with Q235 square steel, while the spacing of steel wire mesh is of 10cm and the diameter of steel wire is of 6mm. Figure 1.b shows an anti-seepage system, including inner water bags and waterproof cloth which use the geomembrane, a widely used anti-seepage material. Compared with rigid water retaining wall using metal plate (steel plate, aluminum alloy plate, etc.), it has good impermeability, strong deformation resistance, good durability and low cost, so it is a kind of excellent impermeable material. Figure. 1.c is the computational model diagram of this study, with bottom two floors in a 1.5m frame structure and the top one in a 1.0m frame structure. The overall layout is staircase piles, and its interior is filled with gravel and water, etc.

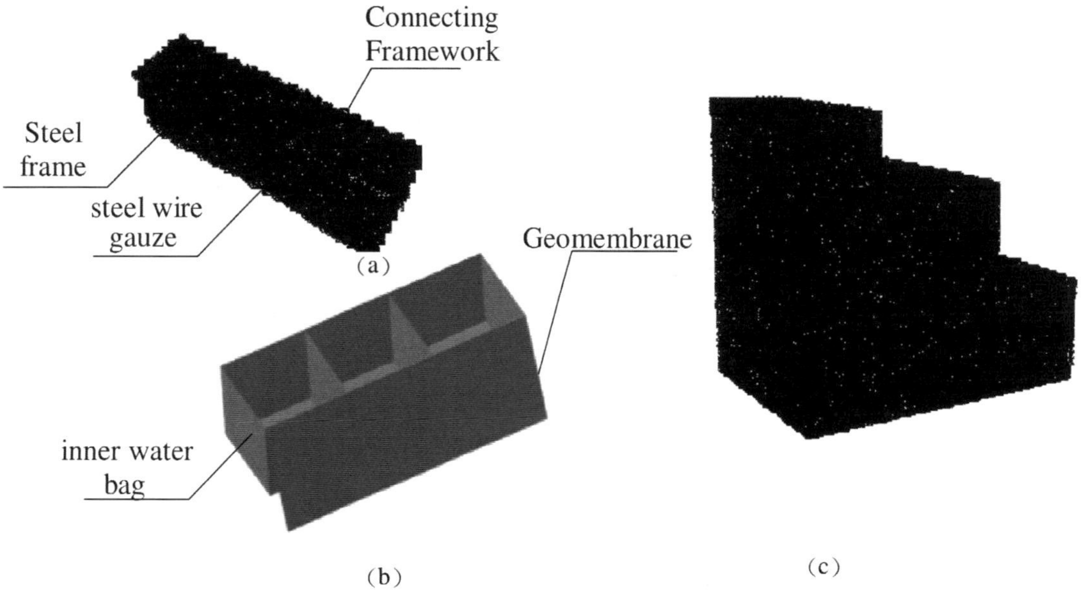

Figure 1. Main Structure of Cage-type flexible Retaining Wall

3. Three-dimensional Finite Element Analysis of Cage-type Retaining Wall

A three-dimensional mesh cage-type flood wall model is established by using finite element analysis software ABAQUS. The size of the frame element of the flood wall is $1 \times 1 \times 1$ m, and the whole flood wall is composed of 7 frame elements. The connection between the frame girder and the beam is made by welding during the test. Therefore, the frame girders can be set as rigid connections in numerical simulation. In the actual structure, the steel wire mesh structure is small in size and large in quantity. Thus, the direct modeling will lead to too large number of meshes, long calculation time, and stress concentration due to inaccurate geometry of the model, which will affect the accuracy of calculation. As a result, the equivalent homogeneous model is adopted in this calculation, which is equivalent by volume to the homogeneous solid model rigidly connected to the main frame structure. The stratum is divided into two layers: the first is loam and the second is bedrock. The calculated length, width and depth of the model take 31m, 37m and 30m respectively, and the calculation model is shown in Figure. 2. Where, Figure.2.a is the overall model diagram, Figure.2.b is the calculation frame model, and Figure. 2.c is the simplified homogeneous model of geomembrane and wire mesh.

There are four boundary types in the calculation model: known head boundary, seepage boundary, anti-seepage boundary and fixed constraint: 1) the known head boundary includes upstream of flood wall as well as the horizontal contact surface between the anti-seepage cloth and water; 2) the seepage boundary is the downstream of flood wall area; 3) the anti-seepage boundary includes the boundary at both sides of flood wall and bottom interception boundary of upstream foundation. The fixed constraint includes the bottom of mode and soil body at two sides. The total number of finite element mesh nodes is 42,883, and the total number of elements is 38,880.

Considering that the main load is from the structural dead weight, external water pressure and internal water pressure, the stress deformation of the structure under normal operation of water retaining are calculated and analyzed. The main calculation parameters are shown in Table 1. According to the water retaining test, it can be known that the flood wall is stable as a whole. Therefore, the whole flood wall is rigidly connected to the foundation in the calculation, with the water retaining height of the flood wall of 1.0m and the downstream water level of 0.0m. The main calculation results are shown in Figure. 3-5.

Table 1. Model Calculation Parameters

Material	Density (kg/m^3)	Permeability coefficient (m/s)	Void ratio	Modulus of elasticity (GPa)	Poisson ratio
Square steel	7800	1.0×10^{-7}	0.2	206	0.26
Wire mesh + Geomembrane	7800	1.0×10^{-6}	0.20	216	0.25
Strata 1	1800	1.0×10^{-7}	0.19	0.05	0.3
Strata 2	1850	1.0×10^{-8}	0.18	0.1	0.23

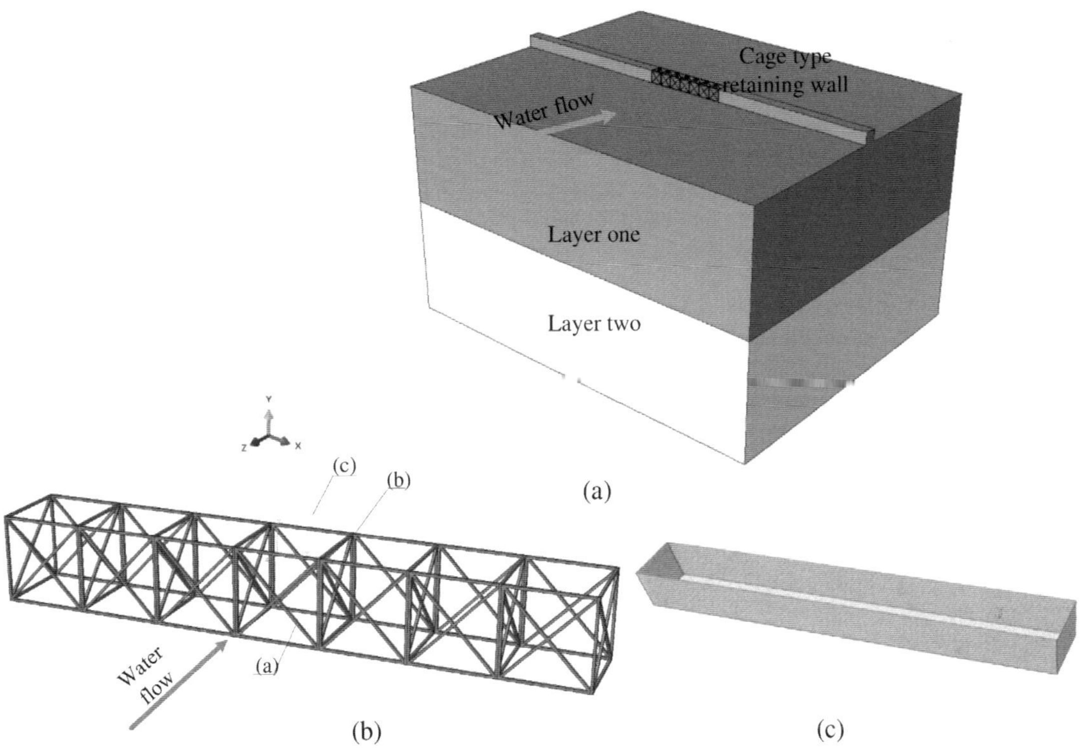

Figure.2. Diagram of Calculation Model

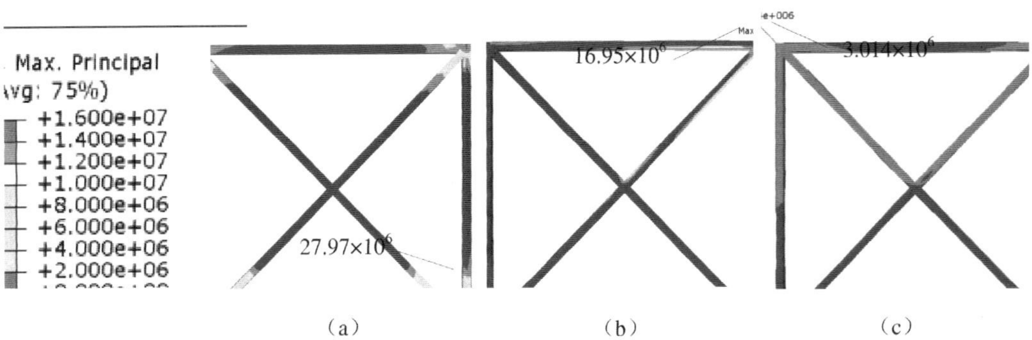

Figure.3. Nephogram for the Large Principal Stress of Main Frame

Figure. 3 is the diagram for the large principal stress of the main frame of the cage-type flood wall. Figure. 3.a is the upstream large principal stress of frame perpendicular to the water flow direction, Figure. 3.b is the large principal stress of frame at the water flow direction, and Figure. 3.c is the downstream large principal stress of frame perpendicular to the water flow direction. In ABAQUS, the tensile stress is taken as the positive stress, so the large principal stress corresponds to the tensile stress. It can be seen from the diagram that the upstream maximum tensile stress of frame perpendicular to the water flow direction appears at the top of the frame intersection, and the maximum tensile stress is 15.9MPa; the maximum tensile stress of frame along the water flow direction appears at the top intersection of frame near the upstream side, and the maximum tensile stress is 3.0MPa; the downstream maximum tensile stress of frame perpendicular to the water flow direction appears at the bottom intersection of the frame, and the maximum tensile stress is 17.1MPa. According to the distribution range of tensile stress over 14MPa, the largest frame in the tension zone of the structure is the downstream vertical flow frame, and the concentration location of tensile stress is located at the junction between the cross beam and the column at the bottom of the frame. The maximum tensile stress of the downstream frame is larger than that of the upstream and down-flow sides, which is because the water pressure inside and outside the upstream face as well as the water pressure on the left and right sides of the frame at the down-flow side are balanced, so this equals to the fact that the frame at the upstream face suffers no pressure. The frame along the down-flow direction only suffers the bottom anchoring force and internal downstream water pressure, while the downstream frame has no balance with the external water pressure, so the tensile area has a large scope and the structural stress form is similar with that of cantilever beam. The maximum tensile stress of the whole frame is 17.1MPa, which is less than the permissible value 215[MPa] of Q235 steel. The tensile strength of the cage-type flexible flood control wall meets the requirements.

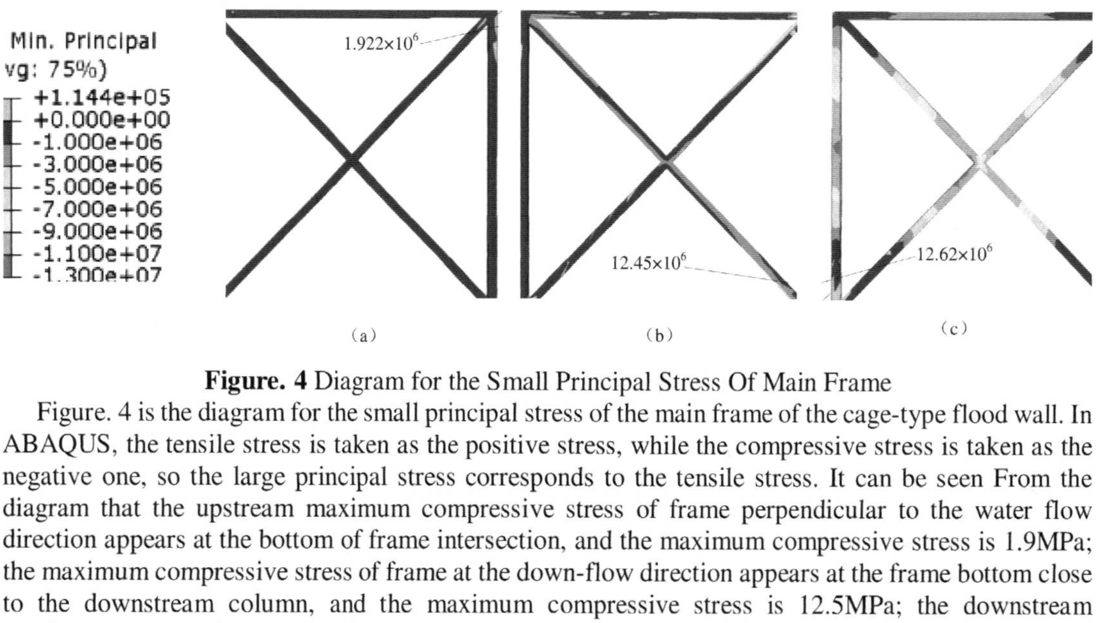

Figure. 4 Diagram for the Small Principal Stress Of Main Frame

Figure. 4 is the diagram for the small principal stress of the main frame of the cage-type flood wall. In ABAQUS, the tensile stress is taken as the positive stress, while the compressive stress is taken as the negative one, so the large principal stress corresponds to the tensile stress. It can be seen From the diagram that the upstream maximum compressive stress of frame perpendicular to the water flow direction appears at the bottom of frame intersection, and the maximum compressive stress is 1.9MPa; the maximum compressive stress of frame at the down-flow direction appears at the frame bottom close to the downstream column, and the maximum compressive stress is 12.5MPa; the downstream maximum compressive stress of frame perpendicular to the water flow direction appears at the bottom of frame column, and the maximum compressive stress is 12.6MPa. According to the distribution range of compressive stress over 10MPa, the largest frame in the structural compression zone is the downstream frame c perpendicular to the water flow and the column bottom of frame b along the water flow, and the concentration part of compressive stress is located at the junction between the bottom cross beam and column of frame. The maximum compressive stress of the whole frame is 12.6MPa, which is less than the allowable value of the tensile stress 215MPa of Q235 steel. The compressive strength of the cage-type flexible flood wall meets the requirements.

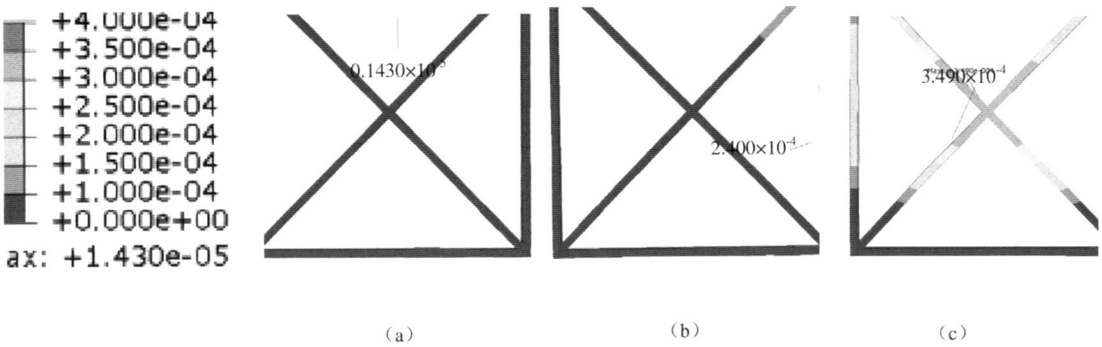

Figure. 5 Diagram for the Deformation of Main Frame

Figure. 5 is the diagram for the deformation of the main frame of the cage-type flood wall. It can be seen from the diagram that the upstream frame a perpendicular to the water flow basically has no deformation, frame b close to the downstream column at the water flow direction has a certain degree of deformation, with the largest deformation of 0.24mm which occurs in the middle of the column. The maximum deformation of the downstream frame c perpendicular to the water flow appears in the middle of the central cross beam, with the maximum deformation of 0.34mm, and there is a certain degree of deformation to the left and right columns. According to the distribution scope of frame with deformation exceeding 0.1mm, the areas with large structural deformation are downstream frame c perpendicular to water flow and the frame b along the water flow direction.

Considering the tensile stress, compressive stress and deformation of the upstream frame a perpendicular to the water flow, frame b along the water flow direction and downstream frame c perpendicular to the water flow, it can be known that the downstream frame c perpendicular to the water flow is the relatively weak part of the structure, while the stress deformation of the upstream frame a perpendicular to the water flow is relatively small. Therefore, when designing the structure, the cross-sectional dimensions of the vertical flow frame on the upstream side and the downstream flow frame can be reduced appropriately, while the cross-sectional dimensions of the vertical flow frame on the backwater side can be increased appropriately, so as to give better play to the material characteristics and save project cost.

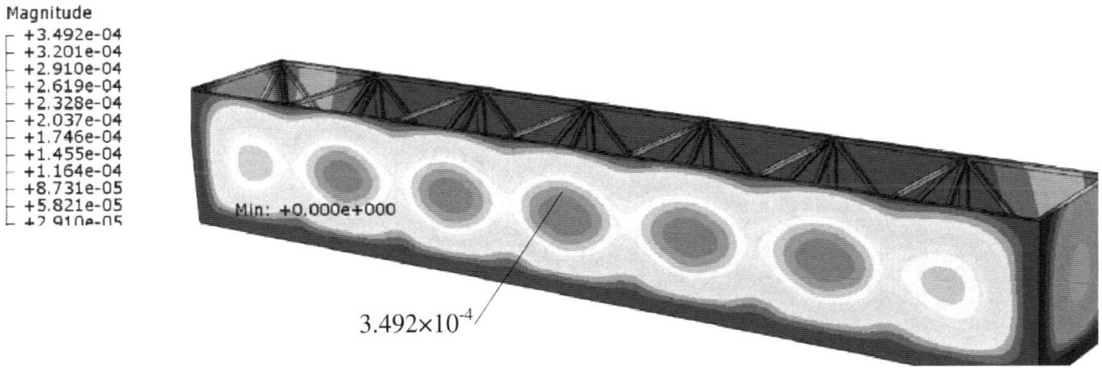

Figure. 6 Diagram for the Deformation of Cage-type Flood Wall Wire mesh

Figure. 6 is the diagram for the deformation of the main frame of the cage-type flood wall. It can be seen from the diagram that the deformation of the mesh mainly concentrates in the downstream middle part, and the maximum deformation is 0.35mm. The deformation of the wire mesh in the middle of the model is larger than that on both sides, which is due to the fixed restraints on both sides of the model itself. The overall structure of the cage-type is equivalent to the beam with fixed restraints at both ends, and the middle part has the largest deformation under the effect of the evenly distributed water pressure. Considering that the geomembrane has 50%-100% ductility, it is certain that the geomembrane will not come up with tensile failure caused by the structural deformation under this deformation condition. The reason why the downstream displacement is larger than the upstream displacement is the same as that of the large principal stress, and it is also because the upstream internal and external water pressure counteracts each other. It can be seen from the deformation of the main frame that the overall deformation of the structure is small, which can meet the requirements of water retaining and flood control.

4. Model Test of Structural Stability and Impermeability

In order to verify the structural stability and impermeability of the new cage-type retaining wall and make sure that it can be used for flood control during the flood season, a full-scale water retaining test of the cage-type retaining wall as the Figure.7 is carried out. A $25 \times 10 \times 3.0$m river course is selected as the test site, and the bottom of the river is a natural river bed without any treatment. After clearing the reeds in the river course, the flexible flood wall is installed. The bottom of the upstream wall is excavated into a slot with the depth of 50cm and width of 20cm. One end of the geomembrane is embedded in the slot, and the other is hung on the top of the cage. Soil bags are used to build bank slopes on both sides of the cage. Geomembrane is used to cover the geomembrane and upstream of the bank slopes to ensure that water will not seep from the junction of the two banks. In this test, 23 frames are set up, the length and height of water retaining is 23m, and the time of water retaining is 30d. The wire mesh is made of 5×7.5mm galvanized low carbon iron wire, the waterproof canvas has an area of 1m^2 and thickness of 3mm, and the steel frame is made of 2×3 cm galvanized steel tube frame.

(a) (b)

Figure.7 Water Retaining Test of The Cage-Type Flexible Retaining Wall

In the course of the river course test, it is found that there is sludge at the bottom of the river course. On the one hand, the sludge reduces the friction coefficient between the bottom of the flood wall and the foundation of the river course due to its soft texture and high fluidity, which leads to the decrease of sliding resistance. On the other hand, the sludge leads to the untight connection between the bottom of water retaining wall and the foundation of river course because of the large permeability coefficient of sludge. At this time, the water in the river course infiltrates into the bottom of the tank body, which increases the uplift pressure of foundation bottom and influences the overall stability of the structure. In order to eliminate the influence of sludge layer below the wall stability, the geotechnical sand bag with its thickness of 3-5cm is laid under the retaining wall. Such sand bags are used to compact the sludge soil by their dead load, so as to increase the friction coefficient between the foundation and the retaining wall, and prevent the excessive uplift pressure of the foundation caused by the lack of close connection. After 30 days of water retaining, there is no obvious seepage behind the wall, which shows that the overall structure has a good anti-seepage effect. The structure remains stable as a whole, and the net cage structure hardly produces deformation, which is consistent with the results of numerical simulation, so the structure can be used for flood control during the flood season. Geotechnical sand bags have good effect in treating sludge foundation and are convenient for construction. It is suggested to adopt geotechnical sand bags according to the actual situation of the project in order to increase the overall stability of the structure.

5. Conclusion

Cage-type flexible retaining wall is a new type of flood control equipment with broad market prospects due to its low cost and convenient installation. In order to study the actual working behavior of cage-type flexible retaining wall during the flood season, a full-scale structural stability and impermeability model test is carried out as well as a three-dimensional finite element structural analysis. The following main conclusions are obtained:

(1) Based on the ABAQUS software, a three-dimensional finite element model of cage-type flood wall is established, and the structural stress deformation analysis under water retaining condition is carried out. The results show that the structural stress and deformation can meet the requirements, the maximum tensile stress is 17.1MPa, the maximum compressive stress is 12.6MPa, and the maximum structural deformation is 0.35mm, which are less than the allowable value and can meet the requirements of engineering construction.

(2) According to the tensile stress, compressive stress and deformation of the frame in different parts, it can be seen that the downstream frame perpendicular to the water flow is the relatively weak part of the structure, while the upstream frame a perpendicular to the water flow has a relatively small stress deformation. In structural design, the cross-sectional dimensions of the upstream frame perpendicular to water flow and the frame at the water flow direction can be reduced appropriately, while the cross-sectional dimension of downstream frame perpendicular to the water flow can be increased appropriately in order to give full play to material characteristics and save project cost.

(3) After 30 days of water retaining, there is no obvious seepage behind the wall, which shows that the overall anti-seepage effect of the structure is good. The overall stability of the structure is maintained, and the net cage structure hardly produces deformation, which indicates that the structure can be used for flood control during the flood season. In practical engineering application, it is suggested that 3-5cm thick sand bags should be used to treat the foundation of the flood wall with sludge, so as to increase the overall stability of the structure.

6. Reference

[1] Wu Junjun, Ni Lijian, Chen Haiyun, Wang Haili. *Experimental Study on the Water Retaining Test of Removable Flood Wall* [J]. International Journal Hydroelectric Energy, 2017,35(03):85-88.

[2] Zhaxi Zhuoma. *Research and Optimal Design Key Technologies of Removable Flood Wall* [D]. North China University of Water Resources and Electric Power, 2018.

[3] Tao Tongkang, Yan Jun, Yu Long. *Thermal Insulation Characteristics of Composite Geomembrane* [J]. Journal of Nanjing Hydraulic Research Institute, 1997(01):11-17.

[4] Shu Yiming. *Discussion on Engineering Characteristics of Impermeable Geomembrane* [J]. Journal of Hohai University, 1993(04):1-6.

[5] Yan Zili, Zhao Leiha, Feng Luming, Wang Yibing. *Review on construction technology and quality control of hydraulic impermeable composite geomembrane* [J]. Yangtze River, 2018,49(S2): 176-181.

[6] Yue Juan, Pan Weizong, An Junjiang, Peng Jing. *Application of Composite Geomembrane in Seepage Control of Yellow River Diversion Gate* [J]. Yangtze River, 2019,41(01):111-115.

[7] Cen Weijun, Wang Hui, Li Dengjn. *Effect of Geomembrane Defects on Seepage Characteristics and Slope Stability of Earth-Rock Dams* [J]. Engineering Journal of Wuhan University, 2018,51(07):589-595.

[8] Duan Yajuan. *Study on Mechanical Properties Test and Construction Technology of Removable Flood Wall* [D]. North China University of Water Resources and Electric Power, 2017.

[9] Guo Lei, Zhaxi Zhuoma, Chen Shoukai, Li Huimin, Wang Lunyan. *Experimental Study on Leakage Characteristics and Mechanical Properties of Flood Control System of the Removable Flood Wall* [J]. Water Resources and Hydropower Engineering, 2018,49(10):80-85.

[10] *Zhao Gengrun, Li Yuwen, Cui Dong, Du Xiaotao. Analysis of Overflowing Risk of Flood Wall in Shanghai Section of Huangpu River [J]. Advances in Science and Technology of Water Resource, 2016,36(03):57-61+67.*

IOP Conf. Series: Materials Science and Engineering **809** (2020) 012014 doi:10.1088/1757-899X/809/1/012014

Automatic Detection Method for Large-Volume Embedded Parts Based on Laser Scanning

Xingyi Hu[1,a], Kaixin Hu[1,b], Hongbo Fan[1,c], Lupeng Yan[1,d], Daguang Han[3,e] Chunli Ying[2,f]

[1]School of Civil Engineering, Chongqing Jiaotong University, Chongqing, China
[2]Academy of Smart City and Sustainable Development, Chongqing, China
[3]Faculty of Technology, Art and Design, Oslo Metropolitan University, Oslo, Norway

[a]huxingyishay@163.com, [b]15823295012@163.com, [c]424048655@qq.com, [d]954518155@qq.com, [e]Daguang.Han@scsda.com.cn, [f]elaineying198429@163.com

Abstract. The embedded part (EMP) is a structural form in a building and plays an important role in the fixation of a structural or a non-structural member. After the initial installation, the popular detection method of the EMPs is to use the total station for single-point diagnosis. Due to the difficulties in point-by-point measurement and feature points selecting, it is sophisticated to ensure accuracy and efficiency in the detection process of large-scale EMPs. In order to solve the above issues, this paper proposes an automatic panel detection method for EMPs: (1) 3D laser scanning is used to obtain the actual coordinates of EMPs; (2) Developed algorithm is made to match the coordinates between design and measured data to calculate construction error; (3) Further use of the optimization algorithm to adjust the attitude of the point cloud, minimize the construction error and give the adjustment guidance of EMPs. A case study indicates that the efficiency is improved by about five times compared to the total station technique, and the number of EMPs need to be adjusted is greatly reduced. This method is of great significance for the high-volume component precision installation project.

1. Introduction

Embedded parts (EMPs) play a critical role in large structures and special buildings (such as laboratory, airport etc.), and their installation accuracy directly determines the safety of the joint framework. The chain reaction caused by the error of the installation position directly gives rise to the structural damage even destruction, which could bring about incalculable injury and estate loss. How to achieve accurate EMPs setting up is a technical problem facing the construction industry. Among them, high-precision detection and arrangement after the preliminary fixing is the key to meeting the construction requirements and ensure the quality.

In this paper, a 3D laser scanner is used to collect 350 sets of EMPs of the Bridge Laboratory in Chongqing Jiaotong University. On this basis, a large number of spatial point clouds are applied for rigid component coordinate, then combined with computer programming. An efficient and automatic batch detection method has been proposed in this paper. Figure.1 shows the research procedure overall. The result of the research demonstrates that it can improves the inspection and adjustment efficiency as well as ensure the accuracy simultaneously.

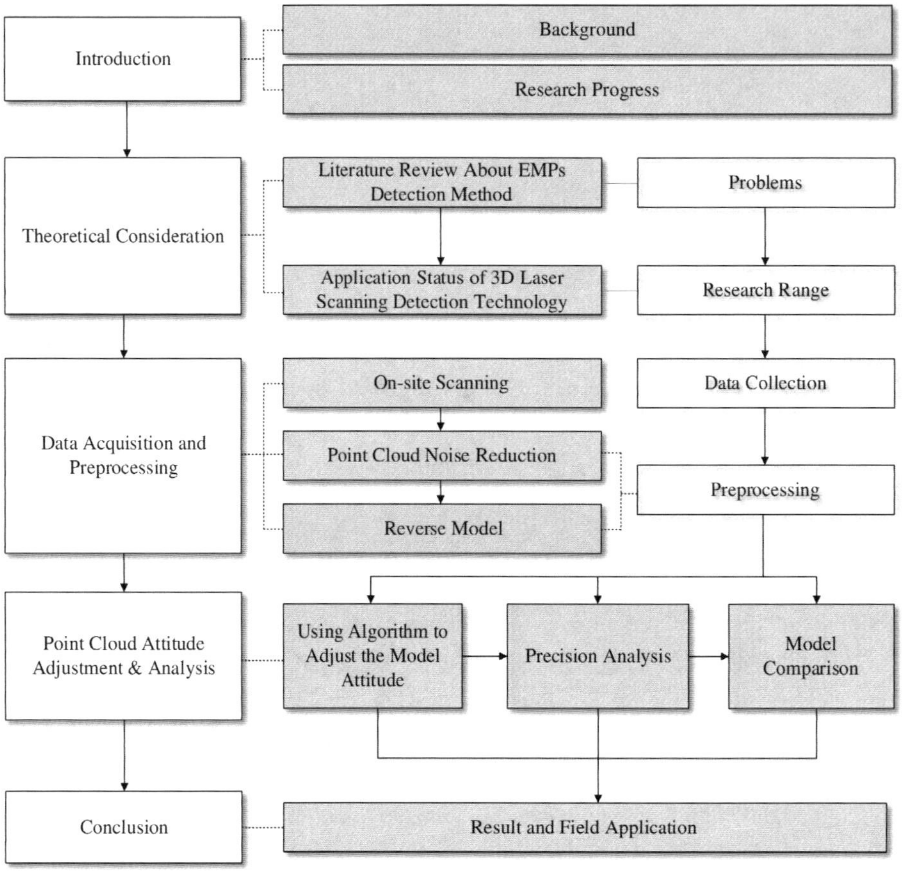

Figure 1. The flow of the research

2. Consideration of the current technologies

Building must be constructed with high precision, which means that the position and elevation of the EMP must meet the requirements. For this purpose, the surveyor is required to measure the spatial position of the EMP before the next component assembling. Unfortunately, this work is time consuming using existent methods, especially when the building is designed with a large number of EMPs. Presently, a total station is of large-scale use to assess the VPOINT one by one, then compare it with the design coordinates. At the meantime, 3D laser scanning is also on the pilot applications in building measurement. Table 1 outlines the pros and cons of above technologies [1,2]. There are a lot of limitation with this traditional method though it is of maturational theory, such as time and labor consuming. Above all it cannot cope with the large quantities of EMPs;

Through the literature review, Li Fengjun [3] established a control network based on the control piles, and adopted the method of erecting fixed brackets to locate the whole EMPs. Although it guarantees the accuracy of the EMPs, there are still some defects: 1) The central point of the EMPs cannot be measured directly, which degrades the positioning accuracy. 2) The EMPs have been detected one by one with total station, the efficiency cannot meet the norm. Tuo Lei [4] introduced a continuous tunnel section interception method based on point cloud, which used a restricted least squares and random sampling consistency algorithm. But it still has not overcome the problem of point cloud stitching accuracy. Li Shen [1] collected point cloud data with 3D laser scanner for a temple by setting the targets masterly, based on which wall thickness, roof depth and other data had been obtained precisely utilizing jointed point clouds inner and outer surface. It is impossible through traditional surveying measures. Glenda K Mayo [5] induced a case study in the application of a historic church disaster, which was documented with the beneficial applications (such as safety and time savings) as well as what the authors believed to be future research areas for similar projects. He also reviews scanning industry applications and some improvements for precision measurements.

Table 1. The pros and cons of different technologies

Technology	Pros	Cons
Total Station	Easy to carry Mature technology Cheaper equipment	Time consuming Lower accuracy Unstructured data Labor intensive Unfit for large-scale areas
3D Laser Scanning	Large amount of data (Accidental error relatively reduced) High precision (sub-millimeter) Full-automatic data collection Visualization	Costly equipment High level technology requirement

In current surveying, the features of the building are analyzed and extracted for detailed surveying with human brain, which gives rise to the subjective measurement result. Furthermore, the unstructured data from traditional methods is difficult to be classified and filed. The advantages of laser scanning, including high accuracy, large volume and visualization, and its numerous applications have made potential solutions to the above problems in modern construction.

3. Methodology

Since the foundation for follow-up structural installation, the accuracy of the EMPs installation determines the safety of the structure directly. For the EMPs' spatial position, the method with 3D laser scanning proposed in this paper contains the following steps:(1) On-site scanning. Due to the complex surrounding at the construction site and the huge volume of EMPs, it is of great significant to select the setting site during data collection. The principles will be covered in section 4.1, which illustrate how to set the measurement system. (2) Point cloud pre-processing. The point cloud collected on site may be separated in multiple parts since the large volume and instrument limitations. It thus requires pre-processing before the next accurate analysis. How to proceed with this step will be elaborated in Section 4.2. (3) Reverse modeling and attitude adjustment. In order to obtain the results of installation deviation of EMPs in the field, this paper establishes the solid model of EMPs with BIM technology based on the pre-processing, which is used for three-dimensional comparison with the design model. (4) Results evaluation. The test results are discussed according to this methodology, as well as the proposed adjustments with a case.

4. Case study

These is a large laboratory under construction in Chongqing Jiaotong University, Chongqing, China (As Figure.2(a) shows. It needs to setup 2,000 sets of counter-force saddles (EMPs) for educational research and experiment. After the installation is completed, it is necessary to ensure whether the EMPs meet the quality requirements: 1) The allowance error of the horizontal distance between the central point of the EMPs roof and the theoretical value is 2mm; 2) the permissible deviation of the verticality is 1/1000 (The tolerance value of the verticality indicates that the angle between the axis of the EMPs and the plane minus 90, the percentage of the result); 3) the allowable deviation of the top plate elevation is 3mm. Figure.2(b) shows the first batch of 350 sets EMPs which needs to be set.

(a) (b)

Figure 2. (a) Bridge laboratory sketch; (b) EMPs (counter-force saddles) installation site

4.1 Onsite scanning.

In order to get higher precision data as much as possible and reduce the work during the post matching, the point cloud collection must comply with some rules. Considering the surrounding on site, the target ball needs to be placed in the appropriate position; at the same time, scanner parameters need to be set in terms of the situation, which includes: pixel pitch, scan rate, and resolution.

Pixel pitch should be determined by the distance between every single station, which must guarantee the minimum gap beneath the allowance error within two different points. Not only the time consuming should be considered, but also the precision needs to be ensured, which decides the scan rate. At the same time, resolution of points is related to the pixel pitch, both of which determined the later process.

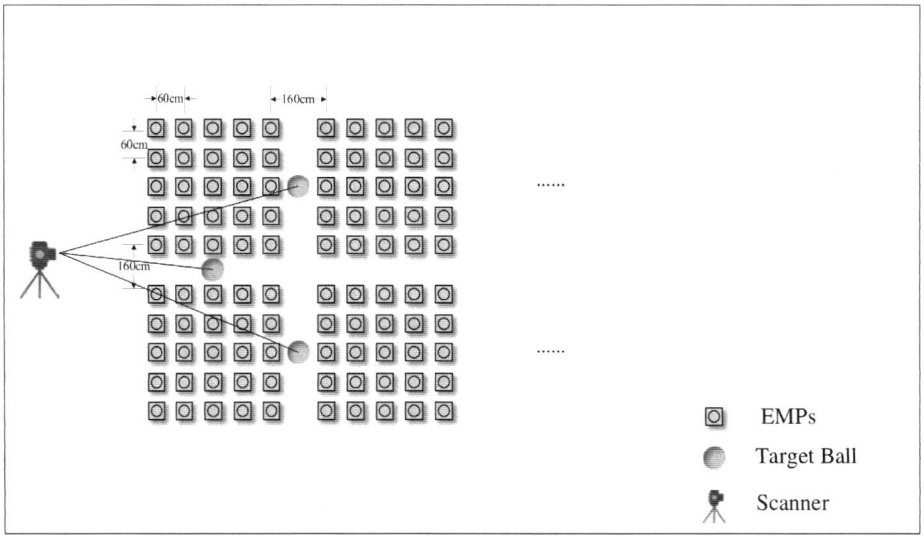

Figure 3. Scanner setup diagram

Figure.3 indicates the layout and circumstances on construction site, FARO X330 3D laser scanner is used in the scanning work. With four target balls keeping stationary between every two stations, the point cloud data with common points is obtained, which provides a basis for the later multi-site cloud splicing. In this whole collection process, six station locations were changed in totally, which brings about six stations of data.

4.2 Point Cloud Pre-Processing

IOP Conf. Series: Materials Science and Engineering **809** (2020) 012014 doi:10.1088/1757-899X/809/1/012014

After point cloud obtained from the above steps, it needs several processing to ensure standard requirements for inspection, including: registration, noise eliminate and converse modelling.

Table 2. Installation deviation analysis data before algorithm processing

Serial Number	Deviation		Serial Number	Deviation		Serial Number	Deviation		Serial Number	Deviation	
	X	Y		X	Y		X	Y		X	Y
3	-0.0005	-0.0026	32	0.0005	0.0021	63	0.0017	0.0023	85	0.0006	0.0023
5	0.0034	0.0010	39	0.0006	0.0038	64	0.0014	0.0027	86	0.0025	0.0019
7	-0.0003	0.0036	40	0.0009	0.0039	65	0.0011	0.0026	87	0.0019	0.0023
8	0.0007	0.0027	43	0.0028	-0.0001	68	-0.0004	0.0024	88	0.0019	0.0035
11	-0.0003	0.0032	48	0.0005	0.0022	69	0.0001	0.0034	89	0.0022	0.0038
13	-0.0012	0.0034	50	0.0030	0.0004	71	0.0021	0.0027	90	0.0007	0.0022
15	0.0000	0.0024	51	0.0008	0.0026	72	0.0008	0.0027	94	0.0011	0.0022
16	0.0004	0.0023	52	0.0016	0.0024	73	-0.0004	0.0031	97	0.0009	0.0028
21	0.0053	0.0004	53	0.0012	0.0024	74	0.0016	0.0032	98	0.0001	0.0026
22	0.0018	0.0035	54	0.0018	0.0026	75	0.0010	0.0022	99	-0.0008	0.0021
25	0.0015	0.0021	55	0.0007	0.0025	76	0.0020	0.0025	100	0.0006	0.0020
28	-0.0001	0.0021	58	-0.0009	0.0021	77	0.0007	0.0025			
29	0.0015	0.0044	61	0.0010	0.0029	78	0.0010	0.0023			
30	0.0005	0.0029	62	0.0001	0.0022	80	0.0009	0.0032			

As it indicates in Figure 4(a), the first thing is to register the six different parts of point cloud. Thanks to the common points on the same target ball between different station, registration can be done by searching for the public surface and points among them using presented means; At the same time, owing to the irregularity of point cloud density and the outliers as well as noise caused by occlusion and other problems, the point clouds need noise reduction. Firstly, unrelated large number of point clouds outside the scene can be deleted manually; then this paper builds a KD-tree for denoising work [6,7] for the disordered point cloud in the field:

(1) Generate a KD-tree according to the point cloud data, and establishing a topological relationship of the point cloud;

(2) Find the neighbor of every point;

(3) Calculate the average distance between the point and each point in the neighbor;

(4) Determine whether the average value exceeds a threshold value: if yes, removed this point.

Figure.4(b) shows the result of reduction. Obviously, the extraneous point cloud has been eliminated. After these processes, the EMPs model could be established from the point cloud. The spatial plane has been constructed below:

$$Ax+By+Cz+D=0 \tag{1}$$

Assuming it is the plane equation of the EMPS' surface as above, and the parameters *A, B, C, D* could be calculated by the principle of least squares method. The space surface has been fitted with the top plate and the cylinder of EMPs' point clouds as Figure.4(c) shows.

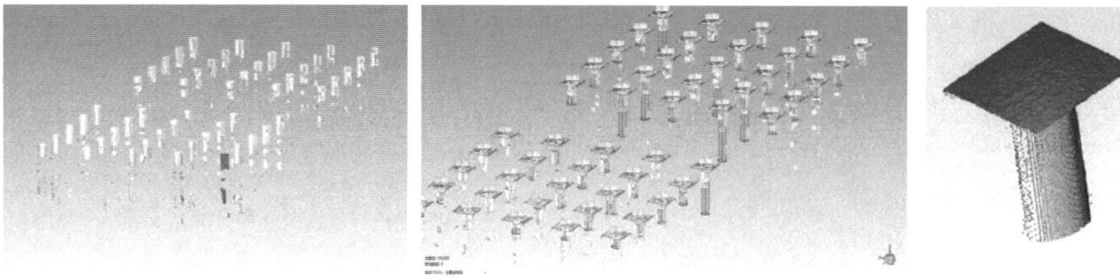

(a) (b) (c)

Figure 4. The point cloud (a) before reduction; (b) after reduction;(c) reverse model

Considering the design data as a reference, the central coordinates and normal equation of the EMPs are extracted from the point cloud. After that, the installation deviation is gained by means of comparing on-site data with reference point clouds. The results are shown in the Table 2 below.

From the chat, it can be concluded that the construction quality does not meet the requirements: The selected 100 sets of EMPs have different degrees of error in X and Y aspects, which means constructor must adjust them to make it below the limitation. However, there will be a mass of work owing to the errors from the test results as well as lacking of exact guidance. What the constructor need emergently is a fast and accurate adjustment method.

4.3 Point cloud attitude adjustment and analyze

In order to solve the issue from what discussed above, the improved ICP algorithm [8-9]is applied for the attitude adjustment. Through rotating and translating the plane, the EMPS' reverse models are corrected as much as possible in an optimized way. According to the algorithm, the coordinate transformation parameters *R* (rotation matrix) and *T* (translation matrix) are obtained through the methodology, so that the distance of 3D data under the two viewing angles is minimized with constantly iteration. Ultimately, the coordinate system is unified in two different point cloud parts. The main steps are as follows:

• Matches the closest point in the reference point cloud (or selected set) for each point in the target point cloud;

• Use Equation 2 to find the rigid body transformation parameters that minimize the root mean square (RMS) of the corresponding point pairs: coordinate transformation parameters *R* (rotation matrix) and *T* (translation vector);

$$\sigma = d(\vec{p_i} + \vec{q_j}) = \sqrt{((x_i - x_j)^2 + (y_i - y_j)^2) + (z_i - z_j)^2} \tag{2}$$

• Use the transformation matrix *R* and *T* to convert the target point cloud;

• Iterate until the condition for terminating the iteration is satisfied (the number of iterations or the error is less than the threshold)

Regarding the calculation of the matrix *R* and *T*, this paper utilizes the classical least squares (LS) principle with MATLAB programming. The LS principle is an optimal estimation technique introduced by the Maximum Likelihood (ML) when the random error is Normal distribution, which minimizes the sum of the squares of the measurement errors.

This paper selects the coordinates of the central point of the EMPs' top plate as the representative. It has been set that: the true central point position coordinate is set *P*, while the ideal central point position coordinate (from design information) is *Q:*

$$P=\{(x_i, y_i), i=1,2,3,\ldots n\}$$

$$Q=\{(x_i'', y_i''), i=1,2,3,\ldots n\} \tag{3}$$

The destination is to achieve the best match, coordinate transformation is employed so that it can align as much as possible between the two set, thus minimizes adjustment of the EMPs.

$$Q=PR+E \tag{4}$$

Among the equation (4), E is the error matrix while R is the rotation matrix:

$$R=\begin{bmatrix}1 & 0 & 0\\ 0 & 1 & 0\\ T_x & T_y & 1\end{bmatrix}\begin{bmatrix}\cos\theta & \sin\theta & 0\\ -\sin\theta & \cos\theta & 0\\ 0 & 0 & 1\end{bmatrix}=\begin{bmatrix}\cos\theta & \sin\theta & 0\\ -\sin\theta & \cos\theta & 0\\ T_x' & T_y' & 1\end{bmatrix} \tag{5}$$

The following equation (6) can be available from formula (5):

$$E=B-AR \tag{6}$$

In order to find the matrix E, the sum of the squares of the two norms of E needs to be minimized:

$$R_m = Min\{E, \|E\|^2 = \|B-AR\|^2\} \tag{7}$$

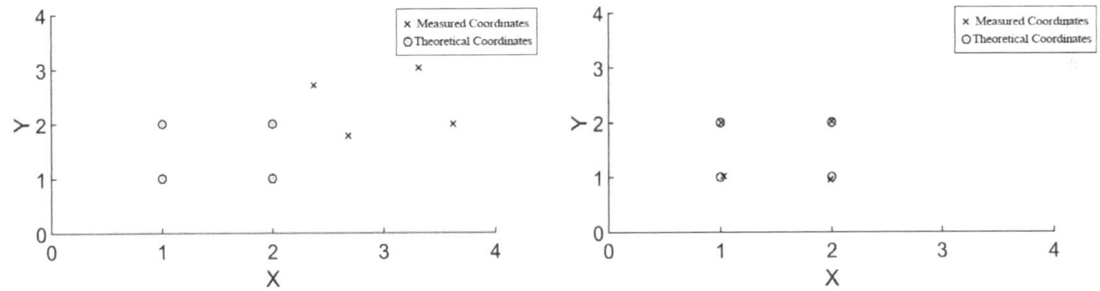

(a) (b)

Figure 5. Using an algorithm to adjust the EMPs, before (a); and after(b)

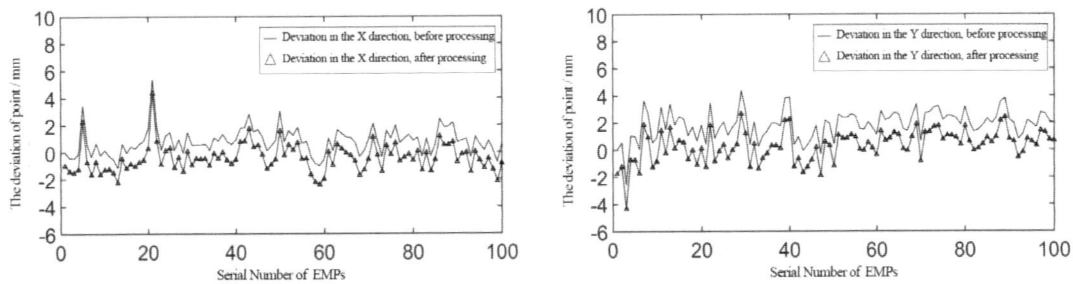

(a) (b)

Figure 6. The deviation after algorithm processing. (a) X direction;(b) Y direction

When E is the smallest, it means R and T are found. Through computer programming, the effect before and after adjustment is as shown in Figure.5. It is obvious that the measured coordinates are coincide with the design values after the algorithm, which indicates that the number of adjusted EMPs can be minimized and the workload can be reduced largely. In fact, all 350 EMPs are adjusted and compared to the design values. The result manifests that the deviation from the design value is significantly reduced, and only 10 EMPs need to be adjusted after using the algorithm (Figure.6).

5. Conclusion

Regardless of whether the algorithm is used, the shape of the deviation curve of the EMPs is similar, which means that the algorithm can only changes the magnitude of the deviation and does not change the shape difference between the whole EMPs' theoretical control network. Both methods are reliable (Figure.6). But, before the algorithm is processed, the number of EMPs needs to be adjusted accounts for 41.4% of the total, and is reduced to 12.3% after being processed by the automated algorithm.

The method proposed by this paper could improve the detection precision of the EMP, greatly reduce the workload in the field, and provide a data reference for the detection and adjustment of the EMP. The use of 3D laser scanning technology and point cloud processing algorithm can greatly improve the efficiency in the civil engineering construction and detection. This will have a profound impact on future detection methods.

Acknowledgments

This research has obtained the financial support of: The Seventh Engineering Co., LTD. of CFHEC, and Academy of Smart City and Sustainable Development in Chongqing. The authors would like to express sincere gratitude to them.

6. References

[1] Shen, L., Tai, H. X., and Ni, T. Y. "Application of 3D laser scanner in surveying of traditional architecture: case study on surveying practice of Hangzhou Haichao temple." *Applied Mechanics and Materials*, pp.353-356, 2013.

[2] Xie, Yuxing, Tian, Jiaojiao and Zhu, Xiao. (2019) A Review of Point Cloud Semantic Segmentation. *IEEE Geoscience and Remote Sensing Magazine*, [Online]. pp. 8-13, Available: https://arxiv.org/pdf/1908.08854v1.pdf

[3] Li Fengjun, Yan Weiwen, Ma Yibin, and Li Cong. "High-precision embedded parts displacement control technology." *Construction technology*, pp.598-600, 2017.

[4] Lei Tuo, Zhizhong Kang, Yuancheng Xie and Baoqian Wang, "Continuously Vertical Section Abstraction for Deformation Monitoring of Subway Tunnel Based on Terrestrial Point Clouds", *Geomatics and Information Science of Wuhan University*, pp.171-175, 2013

[5] Mayo, Glenda. "Point Cloud Applications for Disaster Remediation: A Case Study". *Journal of Facility Management Education and Research*. 2017

[6] Bereczky, Nikolett, Duch, Amalia, Nemeth, Krisztian and Roura, Salvador. (2015). Quad-Kd trees: A general framework for Kd trees and quad trees. *Theoretical Computer Science*. pp. 2015Liu Yanfeng. "Theory and method of point cloud data space management based on Kd-tree." *Doctoral dissertation, Central South University*. 2009

[7] Rusinkiewicz, Szymon. "A symmetric objective function for ICP". *ACM Transactions on Graphics*. 38. pp.1-7. 2019.

[8] R Triebel, W Burgard. "Improving simultaneous mapping and localization in 3D using global constraints". *Proceedings of National Conference on Artificial Intelligence*, January 2005

[9] Marchel, Łukasz and Naus, Krzysztof. "Use of a Weighted ICP Algorithm to Precisely Determine USV Movement Parameters". *Applied Sciences*. pp.9-10, August 2019

IOP Conf. Series: Materials Science and Engineering **809** (2020) 012015 doi:10.1088/1757-899X/809/1/012015

Water Permeability Properties of Concrete made from Recycled Brick Concrete as Coarse Aggregate

Syed Ishtiaq Ahmad[1, a] Md. Shafiqur Rahman[2, b] and Md. Shahrior Alam[2, c]

Department of Civil Engineering, Bangladesh University of Engineering and Technology, Dhaka 1000, Bangladesh

[a]siahmad@ce.buet.ac.bd, [b]shafiqur06@yahoo.com, [c]shahrioralam@yahoo.com

Abstract. Water Permeability characteristic of concrete made from recycled brick concrete as coarse aggregate was investigated and compared with concrete made from crushed virgin clay brick and natural stone aggregate. For this, concrete samples of three different compressive strengths 17.2, 20.6 and 24.1 MPa were prepared as per ACI mix design methods from each of these aggregates. These concrete samples were then tested for compressive strength and water permeability using the AT 315 machine as per BS EN 12390-8: "Depth of Penetration of Water under Pressure". Apart from these, porosity and bulk density were also measured in hardened concrete. Compressive strength of concrete produced from virgin brick and recycled brick concrete was found to be lower than the design strength at 28th day. Coefficient of water permeability of concrete made from recycled brick concrete as coarse aggregate was found to be higher than that of virgin brick and natural stone aggregate concrete of equivalent strength by 350 to 400%. Water permeability was found to be directly related to compressive strength and porosity in hardened concrete.

1. Introduction

Concrete is the principal ingredient in rapidly expanding construction industry of Bangladesh. As there are no major source, natural stone is in short supply and hence expensive in Bangladesh and parts of India. Hence, crushed burnt clay bricks are extensively used as an economic alternative coarse aggregate in construction of midrise residential and factory buildings, rigid pavements as well as small and medium span bridges and culverts [1]. In recent years, many old buildings mainly constructed from brick aggregate concrete are demolished to give away space for new construction. Disposal of demolition waste from these brick aggregate concrete buildings has become a big problem for construction industry. As a consequence, a number of researches are going on examining potential use of recycled brick concrete as coarse aggregate in new concrete production [2]. The durability properties like water permeability, creep and shrinkage has always been a concern for virgin brick aggregate concrete (VBAC) and recycled brick concrete aggregate concrete (RBCAC) [2]. Water permeability is an important parameter regarding durability of concrete. For VBAC and RBCAC, this is even more important issue as both of these aggregates are far more porous and hence permeable than granite and other natural stone aggregate [3-4]. There are a number of works that have been reported till today on durability properties of VBAC and RBCAC [1-8]. Recently, permeability of VBAC has been extensively examined [9]. However, RBCAC is different from VBAC due to presence of micro cracks in recycled brick concrete aggregate and high porosity resulting from those cracks. Recycled aggregate are found to be softer compared to its virgin form. Hence, this is understandable that water permeability would be different in RBCAC than VBAC or natural stone aggregate concrete (SAC). To get comprehensive idea regarding water permeability of RBCAC, an experimental program was conducted at Bangladesh University of Engineering and Technology, Dhaka, Bangladesh. In that project, comparative water permeability behavior of SAC, VBAC and RBCAC were evaluated. For these, concrete samples of three different strength i.e. 17.2 (2500 psi), 20.6 (3000 psi) and 24.1 (3500 psi) MPa were prepared using natural stone, crushed virgin clay brick and recycled brick concrete aggregate. These samples were then subjected to water permeability testing using European standard AT 315 machine as per BS EN 12390-8: "Depth of Penetration of Water under Pressure" [10]. Test results were analyzed to examine the difference in water permeability associated with VBAC and RBCAC. Moreover, compressive strength, porosity and bulk density in hardened concrete were also measured. Influence of these properties on water permeability behavior of corresponding concrete was also investigated.

2. Material Used

2.1 Cement.

Ordinary Portland cement (Type 1) having 28 days compressive strength of 44 MPa as per ASTM C 150 [11] was used for preparation of all concrete samples. By using one type of cement the effect of varying the types of coarse aggregate in concrete was investigated.

2.2 Fine Aggregate.

For maintaining uniformity, one type of natural coarse sand was used throughout the experimental work. Sieve analysis of fine aggregate was carried out in accordance to ASTM C136 [12]. The resulting gradation curve of sand was found to fit within the limits set out in ASTM C33 [13]. Unit weight of fine aggregates was also determined in accordance with ASTM C29 [14] whereas water absorption and specific gravity of fine aggregate was found in accordance with ASTM C128 [15]. From these testing procedures, Fineness Modulus, unit weight, water absorption and specific gravity of fine aggregate were found as 2.55, 1640 Kg/m3, 1.23% and 2.69, respectively.

2.3 Recycled Brick Concrete Aggregate.

In this work, brick aggregate concrete from a recently demolished building was collected. From the collected concrete chunks, cores were extracted and later tested for compressive strength using universal testing machine to assess the quality of brick aggregate concrete. The average recycled concrete core compressive strength was found to be 16.1 MPa. Afterwards, brick concrete chunks were broken by manual labor into small pieces to produce recycled brick concrete aggregate. This sizing of recycled brick was done in a way so that grading limits set out in ASTM C33 [13] could be maintained. Before preparing concrete, different relevant properties of aggregate were measured. This includes bulk density as per ASTM C29 [14], fineness modulus as per ASTM C33 [13], absorption capacity, specific gravity at saturated surface dry (SSD) condition as per ASTM C127 [16] and field moisture content as per ASTM C566 [17]. Test results are reported in Table 1.

2.4 Clay Bricks and stone aggregate.

Virgin Bricks were collected from a brick manufacturing factory located in Dhaka region. Before these bricks were crushed down to aggregate, compressive (crushing) strength test was conducted according to ASTM C 67 [18]. Results of this test show that crushing strength of brick varied between 14 to 29 MPa. Brick aggregate was produced by breaking down whole new bricks on a solid concrete surface using a hammer. Natural sandstone boulders crushed by a stone crusher were used as stone aggregate. Size of both brick and natural stone aggregate were maintained in such a way that that grading limits set out in ASTM C33 [13] was strictly maintained. All relevant properties of brick and natural stone aggregate that include bulk density, fineness modulus, absorption capacity, specific gravity and field moisture content were measured as per relevant ASTM standards and are reported in Table 1. As can be seen from Table 1, absorption capacity of virgin brick and recycled brick concrete aggregate are much higher than natural stone aggregate concrete. Specific gravity at SSD condition for natural stone aggregate is also approximately 20 to 22% higher than brick and recycled brick concrete aggregate.

3. Testing Scheme

3.1 Mix Design and Mixing Method.
A total of nine mix design as per ACI 211 [19] were prepared to conduct the experimental program. Three mixes were made from each of the aggregate type to produce SAC, RBAC and VBAC for target strength of 17.2, 20.6 and 24.1 MPa considering water cement ratio of 0.75, 0.69 and 0.63, respectively. Mix design ratios are shown in Table 2. Since water absorption of virgin brick and recycled brick concrete aggregate was much higher, these aggregates were submerged in water for 48 hour and later wiped to achieve SSD condition prior to adding to the concrete mixture [2, 3, 6]. Otherwise, a large part of water from mix design calculation would had soaked by these aggregates and would not be available to react with cement changing the water cement ratio of the entire mix. This procedure was followed for natural stone aggregate as well. Water absorbed in the aggregate was in addition to the water requirement from mix design calculation as shown in Table 2. For each mix design, nine cylinders (100mm x 200mm) were prepared for compressive strength test at 3, 7 and 28 days. Three cubes (100mm x 100mm x 100mm) were also prepared for each of the mix to measure density, water absorption, and voids in the hardened concrete. Additionally, for each mix design, three cubes (150mm x 150mm x 150mm) were prepared for permeability test.

Table 1. Properties of Different Aggregate

Aggregate Type	Bulk density (kg/m^3)	Fineness Modulus	Specific gravity (SSD)	Absorption capacity (%)	Field Moisture content (%)
Natural Stone	1591	7.42	2.67	0.5	1.8
Virgin Brick	1115	7.29	2.10	8.96	17.96
Recycled Brick Concrete	1101	7.25	2.04	11.24	17.77

Table 2. Concrete Mix Design (Quantity in kg for single batch)

	Strength (MPa)	w/c	Cement	Coarse Aggregate	Fine aggregate	Water
Natural Stone	17.2	0.75	16.9	56.3	56.3	12.7
	20.6	0.69	18.5	56.3	55.0	12.7
	24.1	0.63	20.2	56.3	53.6	12.7
Brick	17.2	0.75	16.9	70.8	57.0	12.7
	20.6	0.69	18.5	70.8	55.7	12.7
	24.1	0.63	20.2	70.8	54.3	12.7
Recycled Brick Concrete	17.2	0.75	16.9	56.5	54.0	12.7
	20.6	0.69	18.5	56.5	52.7	12.7
	24.1	0.63	20.2	56.5	51.3	12.7

3.2 Water Permeability Testing.
European standard AT 315 apparatus was used to determine the water permeability of concrete according to BS EN 12390-8 [10]. The apparatus was connected to a normal air compressor capable of ensuring at least 5 bar compressed air continuously and equipped with dehumidifier and oil filter. Connection then was made to the laboratory water supply and to a drainage system. A specimen was subjected to test when its age was at least 28 days. For testing, the specimen was placed on the apparatus in such a manner that the

water pressure act on the test area which actually is a 75mm diameter area at the center of the bottom surface of the cube sample. Water pressure of (500 ± 50) kPa for (72 ± 2) hours was applied on this surface. After the pressure had been applied for the specified time, the specimen was removed from the apparatus. The face on which the water pressure was applied was wiped to remove excess water. The specimen was then split in half, perpendicularly to the face on which the water pressure was applied. As soon as the split face has dried to such an extent that the water penetration front could be clearly seen, maximum depth of penetration under the test area was recorded and measured to the nearest millimeter. Figure 1 shows example of such penetration area and marked penetration front in a brick aggregate concrete sample.

The depth of water penetration inside the specimen can be converted to its equivalent coefficient of water permeability using Valenta's equation [20]:

Figure 1. Water penetration area and penetration front in a brick aggregate concrete sample

$$k = \frac{e^2}{2ht} v \text{ m/s} \quad (1)$$

Where, e= depth of penetration of concrete in meters; h= hydraulic head in meters; t= time under pressure in seconds, and v= the fraction of the volume of concrete occupied by pores. The value v represents discrete pores, such as air bubbles, which do not become filed with water except under pressure, and can be calculated from the increase in the mass of concrete during the test.

4. Results and Discussion

4.1 Compressive strength of concrete.

Strength of concrete samples were measured through crushing of cylinder specimens using universal testing machine. Figure. 2(a), 2(b) and 2(c) show compressive strength that was observed for three different design strength i.e. 17.2, 20.6 and 24.1 MPa respectively for SAC, VBAC and RBCAC. For all sample type, compressive strength was measured at the age of 3, 7 and 28 days. Observation of Figure. 2 (a), (b) and (c) shows that at 28th day, design strength could not be achieved for VBAC and RBCAC. VBAC compressive strength at 28th day was approximately 7% lower the design compressive strength. Whereas, for RBCAC, measured compressive strength at 28th day was found to be approximately 11% lower than the corresponding target strength. This is in accordance to the findings of the previous researchers, where use of brick and recycled aggregate achieved lower compressive strength compared to the target strength [5,6,7]. Additionally, from these results, it apparent that compressive strength of concrete produced using virgin brick aggregate was slightly higher than that of concrete produced using recycled brick concrete as coarse aggregate keeping all other parameters constant. Observation of Figure. 2(a), (b) and (c) also indicate that at early age i.e. 3rd day after casting, VBAC shows slightly higher compressive strength than SAC for 17.2 and 20.6MPa target strength concrete. However, rate of strength gain decreased for VBAC and RBCAC compared.

4.1.1 Porosity, absorption and bulk density of concrete.

Porosity, absorption and bulk density of concrete are important parameters that affect its permeability. For all types of concrete in this work, these parameters were measured as per ASTM 642 [21]. Relationship between design strength and porosity, absorption and bulk density of concrete made from different aggregates are shown in Figure.3(a), 3(b) and 3(c), respectively. As can be seen from these figures, for equivalent design compressive strength, orosity and water absorption in RBCAC is approximately twice of that found in SAC. VBAC has higher porosity and water absorption than SAC but much lower than RBCAC. Porosity in VBAC was found to be approximately 35% higher than SAC. Bulk density of RBCAC was found to be approximately 15% lower than SAC of corresponding strength. Bulk density of VBAC was found to be slightly higher than RBCAC but much lower than SAC. Both VBAC and RBCAC possess much higher porosity and absorption than SAC.

4.1.2 Coefficient of Permeability for different concrete.

Coefficient of permeability for three different types of concrete was measured as per BS EN 12390-8 [10]. Figure. 4(a) shows coefficient of permeability for SAC, VBAC and RBAC with respect to design strength in concrete. As can be seen from these figures, coefficient of permeability decreased with increase in design strength. For concrete with comparatively lower design strength, i.e. 17.2 MPa, coefficient of permeability in

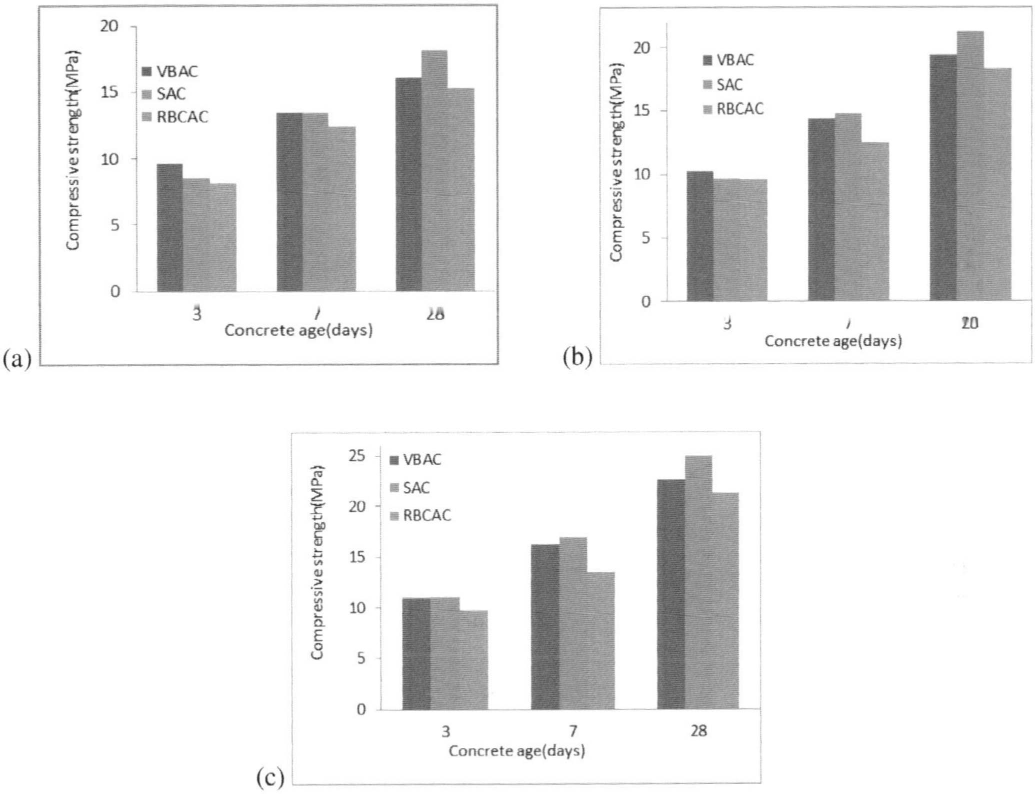

Figure.2 Compressive strength comparisons with the age of concrete for (a) 17.2 MPa (b) 20.6 MPa and (c) 24.1 MPa design strength.

VBAC is approximately twice the value of SAC. For this design strength in RBCAC, coefficient of permeability was even higher and found to be approximately 300% of corresponding value in SAC. However, as design compressive strength was increased from 17.2 to 24.1 MPa, coefficient of permeability in VBAC decreased rapidly and approached a value similar to SAC. Coefficient of permeability in RBCAC also decreased with increased compressive strength but still showed much higher value than VBAC and SAC. Coefficient of permeability in RBCAC was found to be 350 to 400% higher than corresponding SAC of equivalent compressive strength. And, coefficient of permeability was found to be higher than VBAC by approximately 200%. Hence, RBCAC should be used with extreme caution. Figure.

4(b) shows variation of coefficient of permeability with respect to porosity of concrete. Explicably, increase in porosity of concrete was associated with increased coefficient of permeability.

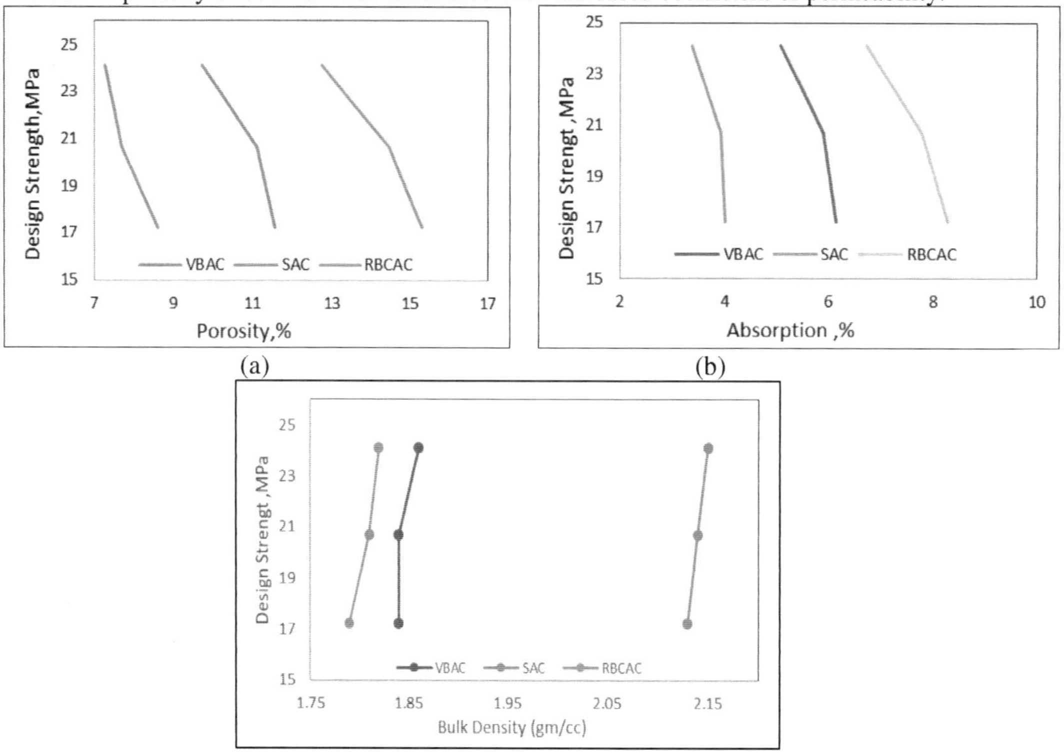

Figure. 3 Relationship between compressive strength and (a) Porosity (b) Absorption and (c) Bulk Density for concrete made from different aggregates

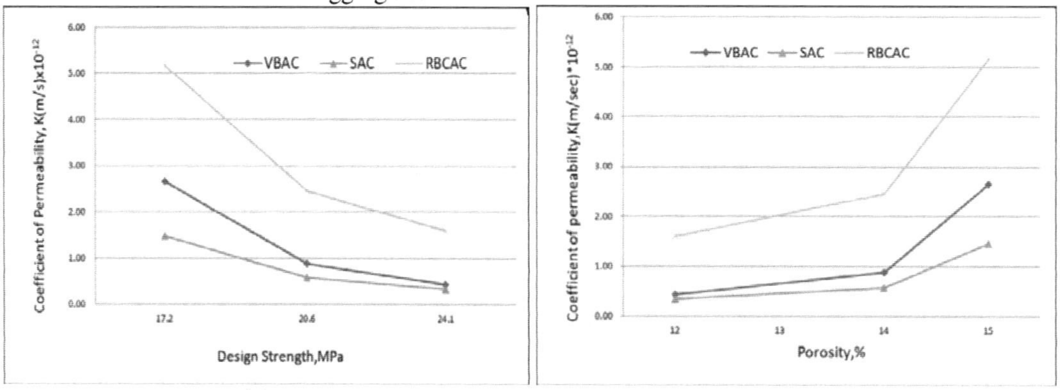

(a) (b)

Figure. 4 Relationship between Coefficient of Permeability and (a) Design strength (b) Porosity for concrete made from different aggregates

5. Conclusion

Coefficient of permeability for three different types of concrete having three different coarse aggregate, i.e. natural stone, virgin crushed clay brick and recycled brick concrete were measured in this work. For this concrete samples were prepared and different tests were conducted on those samples to measure compressive strength, porosity, water absorption, bulk density and coefficient of permeability. It was found that design compressive strength as per ACI mix design could not be achieved at 28th day for VBAC and RBCAC Porosity and water absorption was found to be almost twice in RBCAC with respect to corresponding value for SAC. It has also been found that coefficient of permeability increased with decrease in design strength and vice versa for all types of concrete. Additionally, it has been observed that coefficient of permeability in RBCAC is 350 to 400% higher than SAC of corresponding compressive strength. Coefficient of permeability was also found to be much higher than VBAC. Hence RBCAC should be used with care and may not be used for important elements of structures or buildings without implementing appropriate precaution against permeability related durability problems.

6. References

[1]M. A. Mansur, T.H. Wee, and S.C. Lee, "Crushed Bricks as Coarse Aggregate for Concrete," ACI Mater. J., vol. 96, pp. 478-484, 1996.

[2]Mohammed, T.U., Hasnat, A, and Awal, M.A., "Recycling of Brick Aggregate Concrete as Coarse Aggregate", J. Mater. Civ. Eng., vol. 27, no. 7, 2015.

[3]F.M. Khalaf, "Using Crushed Clay Brick as Coarse Aggregate in Concrete," J. Mater. Civ. Eng., vol. 18, no. 4, pp. 518-526, 2015.

[4] A.A. Akhataruzzaman, and A. Hasnat, "Properties of Concrete Using Crushed Brick as Aggregate," ACI Concr. Int., vol. 5 pp. 58-63, 1983.

[5]P.B. Cachim, "Mechanical Properties of Brick Aggregate Concrete,"Constr. Build. Mater., vol. 23, pp.886-893, 2009.

[6] S.I. Ahmad and S. Roy, "Creep Behavior and Its Prediction for Normal Strength Concrete Made from Crushed Clay Bricks as Coarse Aggregate," J. of Mater. Civil Eng., Vol. 24, pp. 308-314, 2012.

[7] S.I. Ahmad and S. Roy, "Creep Prediction Model for Concrete made from Crushed Clay Bricks as Coarse Aggregate," Applied Mechanics and Materials, v. 166-169, pp. 994-997, 2012.

[8] M. Adamson, A. Razmjoo, and A. Poursaee, "Durability of Concrete Incorporating Crushed Brick as Coarse Aggregate," Constr. Build. Mater., vol. 94, pp. 426-432, 2015.

[9] S.I. Ahmad and M. A. Hossain, "Water Permeability Characteristics of Normal Strength Concrete Made from Crushed Clay Bricks as Coarse Aggregate," Advances in Material Science and Engineering, vol. 2017, Article ID 7279138, 2017.

[10] BSI, "Testing Hardened Concrete – Part 8: Depth of penetration of water under pressure," BS EN 12390-8, Chiswick High Road, London, 2009.

[11] ASTM,"Specification for Portland cement," ASTM C 150, West Conshohocken, PA, 2007.

[12] ASTM, "Standard test method for sieve analysis of fine aggregate," ASTM C 136, West Conshohocken, PA, 2014.

[13] ASTM, "Standard Specification for Concrete Aggregates," ASTM C 33, West Conshohocken, PA, 2011.

[14] ASTM, "Standard Test Method for Bulk Density ("Unit Weight") and Voids in Aggregate," ASTM C29, West Conshohocken, PA, 2009.

[15] ASTM, "Standard Test Method for Density, Relative Density (Specific Gravity), and Absorption of Fine Aggregate," ASTM C 128, West Conshohocken, PA, 2015.

[16] ASTM, "Standard Test Method for Relative Density (Specific Gravity) and Absorption of Coarse Aggregate," ASTM C 128, West Conshohocken, PA, 2015.

[17] ASTM, "Standard Test Method for Total Evaporable Moisture Content of Aggregate by Drying aggregate," ASTM C 566, West Conshohocken, PA, 2019.

[18] ASTM," Standard Test Methods for Sampling and Testing Brick and Structural Clay Tile," ASTM C67, West Conshohocken, PA, 2014.

[19] American Concrete Institute (ACI), "Standard Practice for Selecting Proportions for Normal, Heavyweight and Mass Concrete," ACI 211.1-91, Famington Hills, MI, 2002.

[20] O. Valenta, "The Permeability of Concrete in Aggressive Condition," Proceedings of the 10th International Congress on Large Dams, pp.103-117, 1970.

[21] ASTM, "Standard Test Method for Density, Absorption and Voids in Hardened Concrete," ASTM C 642, West Conshohocken, PA, 2013.

Investigation of the Significance of Factors Affecting Long-Term Creep of Concrete using Screening Designs

H Daou[1] and W Raphael[1]

[1] Saint-Joseph University, Beirut, Lebanon

Email: hikmat.daou@net.usj.edu.lb

Abstract. Creep has a main role in predicting delayed strain in concrete members and in estimating losses in prestressed concrete, such as nuclear containment vessels, bridges girders, etc. For that, the interest in studying creep has grown significantly. This study aims to investigate the significance of different factors affecting creep of concrete using the Northwestern University (NU) creep database. Experiments have been carried out at different conditions of water-cement ratio (w/c), aggregate-cement ratio (a/c), cement content (c), cement type (R or RS), admixtures, mean compressive strength (f_{cm}), volume-surface ratio (v/s), relative humidity (RH), age at loading (t_0) and stress level (sigma). In order to determine the significance of these factors, three statistical methods have been used: Daniel Plot, Lenth Plot and Bayesian screening. The results of this analysis can be used to build accurate computational models to predict creep of concrete by considering the major effects identified.

1. Introduction

Concrete is known by its time-dependent volume changes where it shrinks during drying and creeps under sustained stresses. These changes in volume lead to an excessive long-term deflection of the structural members, loss of prestressing force in prestressed concrete elements, wide cracks in the tension members and redistribution of stresses with time in composite concrete structures. Therefore, designers and engineers must accurately predict creep and shrinkage strains in their structural analysis. Creep strain represents the time-dependent increase in strain under sustained constant load taking place after the initial strain at loading. The creep strain may be several times greater than the initial strain. Creep strain may be subdivided into a drying and a non-drying component, termed drying and basic creep, respectively. Basic creep represents the creep at constant moisture content with no moisture movement through the material, and is consequently independent of the specimen size and shape. It has not been determined whether basic creep approaches a final value, even after 30 years of measurement of sealed specimens [1, 2]. Drying creep is the additional creep occurring in a specimen exposed to the environment and allowed to dry. As it is caused by the drying process, drying creep depends on the size and shape of the specimen and may be expected to show a limiting value at long term [3, 4].

Creep mechanism is a complex process that is affected by significant number of parameters. For this purpose, it is crucial to know these parameters affecting creep of concrete. Many studies found in literature study the factors affecting the creep of concrete: quantity of aggregate [5], properties of aggregate [6], water content, cement content [7], air content [8], type of cement [9], relative humidity [10], temperature [11, 12], load [13], curing [14], size and shape of members [15] and loading age [16]. The creep of concrete is also affected by the admixtures: for example, the concrete containing silica fume exhibits a lower creep than the controlled (no silica fume) specimen under drying conditions while it showed a 12% higher creep than the controlled specimen under sealed conditions [17]. Moreover, Mazloom et al. [18] shows that the creep strain decreased with increase in silica fume. Some studies of the effect of the fly ash shows that the increase in fly ash content increases creep [19, 20] where others studies conclude that the increase in fly ash content decreases the creep [21, 22]. Other authors studied the effects in self compacted concrete mixes containing mineral admixtures [23 - 25].

The prediction of creep strain in concrete structures is important to assess its serviceability and durability conditions. Many models for predicting creep are proposed for both normal and high strength concrete [26-33]. The model recommended by ACI Committee 209 found in ACI 209R-92 [34] to predict creep requires age of concrete when drying starts, age of concrete at loading, curing method, relative humidity, volume-surface ratio and cement type. The Bazant-Baweja B3 model [35] clearly separates basic and drying creep and considers the following factors to predict creep: age of concrete when drying starts, age of concrete at loading, aggregate content, cement type, mean concrete compressive strength at 28 days, curing method, relative humidity, shape of specimen, volume-surface ratio and water content in cement. CEB MC90-99 model [36] presents the prediction model of shrinkage and creep in concrete for both normal and high-strength concrete. This model requires age of concrete when drying starts, age of concrete at loading, mean concrete compressive strength at 28 days, relative humidity, volume-surface ratio and cement type. The GL2000 model [37] is also a model used to predict shrinkage and creep in concrete requiring age of concrete when drying starts, age of concrete at loading, relative humidity, volume-surface ratio, cement type and mean concrete compressive strength at 28 days. The factors required in creep model prediction should be the most factors affecting creep of concrete. The previous studies investigate the factors affecting creep of concrete by changing one factor and keeping all others factors as constant. Therefore, the weight of effect or the significance of factor cannot be assessed.

This study aims to investigate the significance of ten factors affecting the creep of concrete: water-cement ratio, aggregate-cement ratio, cement content, mean concrete compressive strength, effective thickness, age at loading, relative humidity, sustained stress, type of cement, and presence of admixtures. The NU database is used and the analysis is carried out using different methods of screening designs: Daniel Plot, Lenth's Plot and Bayesian screening. The results can be used to build accurate computational models to predict creep of concrete by considering the major effects identified in this analysis.

2. Experimental database

This study used the latest database assembled at Northwestern University (NU) during 2010-2013 under the support of the U.S. Department of Transportation [38]. This database is constituted of approximately 1433 creep tests, of which approximately 800 creep curves contain admixtures. The tests in this database were performed using different parameters and under various conditions. For each test, the mix proportions, testing conditions, specimen geometries and admixtures are documented and the following parameters are provided: water-cement ratio, aggregate-cement ratio, cement type , mean concrete compressive strength, effective thickness, age at loading, temperature, relative humidity, sustained stress, and admixtures type and percentage.

This study aims to study the significance of factors affecting long-term creep; therefore the data used was limited for creep compliance at 3000 days and ten factors were considered as independent variable: water-cement ratio (w/c), aggregate-cement ratio (a/c), cement content (c in kg/m^3), mean concrete compressive strength (f_{cm} in MPa), effective thickness (v/s in mm), age at loading (t_0 in days), relative humidity (RH in %), sustained stress (sigma in MPa), type of cement (R or RS) and presence of admixtures. In order to conduct screening designs, the data used in this study was organized and converted to a two-level factorial design as shown in Table 1.

IOP Conf. Series: Materials Science and Engineering **809** (2020) 012016 doi:10.1088/1757-899X/809/1/012016

3. Methodology

Table 1. Division criteria of data.

Factor	-1	+1
w/c	< 0.5	≥ 0.5
a/c	< 5	≥ 5
c	< 350	≥ 350
f_{cm}	< 50	≥ 50
t_0	< 28	≥ 28
RH	< 55	≥ 55
v/s	< 30	≥ 30
Sigma/f_c	< 0.3	≥ 0.3
Type of cement	R	RS
Existence of admixture	no	yes

In order to study the significance of the factors affecting long-term creep of concrete, the concrete compliance function $J(t,t_0)$, defined using Equation (1), was unified for t=3000 days.

$$J(t,t_0) = \frac{1}{\sigma(t_0)}[\varepsilon_{el}(t_0) + \varepsilon_{cr}(t,t_0)] \tag{1}$$

where $\sigma(t_0)$ is the sustained stress, $\varepsilon_{el}(t_0)$ is the instantaneous elastic strain, and $\varepsilon_{cr}(t,t_0)$ is the creep strain between t, time of evaluating creep, and t_0, time of load application.
In the NU database, there are many tests where readings were stopped after a period of loading that is less than 3000 days while other tests have readings that exceed 3000 days with no reading at this time. For that, extrapolation (see Equation (2)) and interpolation (see Equation (3)) were applied to calculate $J_{exp}(3000,t_0)$ using the compliance J_{EC2} calculated using the equation of the Eurocode 2 [39].

$$J_{exp}(3000,t_0) = J_{exp}(t_1,t_0) + [J_{EC2}(3000,t_0) - J_{EC2}(t_1,t_0)] \tag{2}$$

$$J_{exp}(3000,t_0) = J_{exp}(t_1,t_0) + [J_{EC2}(3000,t_0) - J_{EC2}(t_1,t_0)] \times \frac{J_{exp}(t_2,t_0) - J_{exp}(t_1,t_0)}{J_{EC2}(t_2,t_0) - J_{EC2}(t_1,t_0)} \tag{3}$$

where $J_{exp}(t_1,t_0)$ and $J_{exp}(t_2,t_0)$ are the results of the compliance at time t_1 and t_2, respectively with t_1 days < 3000 days < t_2 days; $J_{EC2}(t_1,t_0)$, $J_{EC2}(3000,t_0)$ and $J_{EC2}(t_2,t_0)$ are the compliance predicted by Eurocode 2 at time t_1, 3000 and t_2 days, respectively.
In screening experiments, factor sparsity is usually assumed. It means only a few of all factors considered in the experiment will actually affect the response. The statistical methods used in this study are: Daniel Plot, Lenth's Plot and Bayesian screening.

3.1. Daniel plots

Daniel Plots, which is also known as normal plot of effects, arrange the estimated factors effects in a normal probability plot [40]. The factors "out of the straight line" are identified as potentially active factors. It is an important, commonly used and successfully employed tool for identifying important factors in 2-level full and fractional experiments.

The normal probability plot of the effects shows the standardized effects relative to a distribution fit line for the case when all the effects are 0. The standardized effects are t-statistics that test the null hypothesis that the effect is 0. Positive effects increase the response when the settings change from the low value of the factor to the high value. Negative effects decrease the response when they settings change from the low value of the factor to the high value of the factor.

3.2. Lenth's plots

Lemth's method is used to assess the effect of factors and based on factor sparsity. This method defines the pseudo standard error (PSE) (see Equation (4)) and the 95% margin of error (ME) (see Equation (5)).

$$PSE = 1.5 \times \underset{|c_j|<2.5s_0}{median} |c_j| \tag{4}$$

$$ME = t_{0.975,d} \times PSE \tag{5}$$

where $t_{0.975,d}$ is the 0.975^{th} quantile of the t distribution with $d = \frac{m}{3}$ degrees of freedom, $c_1,\ldots,c_m$ are the estimated contrasts and $s_0 = 1.5 \times median|c_i|$, is the approximate standard error.

Finally, the 95% simultaneous margin of error (SME) (see Equation (6)) is defined for simultaneous inference on all the contrasts.

$$SME = t_{\gamma,d} \times PSE \tag{6}$$

where $\gamma = 0.5(1 + 0.95^{-m})$ [41].

3.3. Bayesian screening

Bayesian screening is based on the factor sparsity hypothesis. For the linear model $y = X\beta + \varepsilon$, the procedure assigns prior normal distributions $N(0,\gamma^2\sigma^2)$ to each of the independent β_i, where γ^2 is the magnitude of the effect relative to the experimental noise and σ^2 is the variance of the error. The assumption of the factor sparsity is manifested by assigning a prior probability π to any factor of being active and $(1-\pi)$ to the factor of being inactive. The posterior probabilities of the models M_i for all subsets of factors are then calculated. Consequently, marginal factors posterior probabilities p_i are computed. Finally, the factor effects with higher probabilities are identified as potentially active [42, 43].

4. Results and discussion

The aim of this study is to determine the significant factors affecting creep of concrete. For that, a screening analysis of two-level factorial design was carried out and examined ten factors affecting concrete compliance. Figure 1 shows the half-normal plot of effects (Daniel Plot). In normal and half-normal plots, the factors "out of the straight line" are identified as potentially active factors; therefore, as seen in Figure 1, it is not clear which factors that may be active.

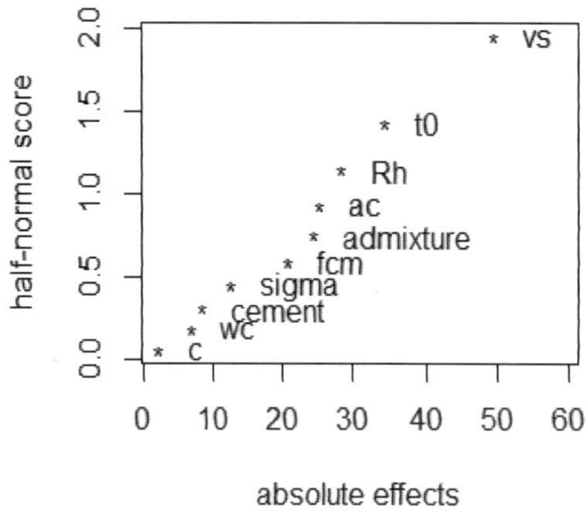

Figure 1. Half-normal plot of effects.

Lenth's plot is presented in Figure 2 and as seen none of the effects goes beyond the margin of error ME, thus the SME limits are not displayed, so the identification of the active factors is not possible.

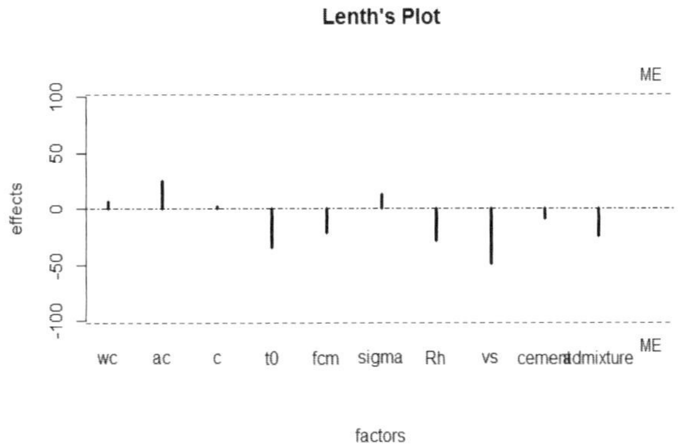

Figure 2. Lenth's plot.

Figure 3 shows the results of Bayesian screening. The most individual factors affecting creep of concrete are ordered as follows: the mean compressive strength, the volume-surface ratio, the aggregate-cement ratio, the water-cement ratio, the presence of admixtures, the time of first loading, the content of cement, the cement type, the relative humidity and the sustained load.

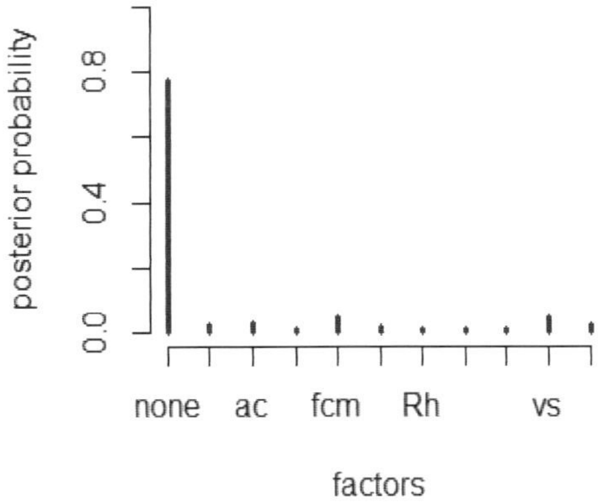

Figure 3. Posterior probabilities of factors using Bayesian screening.

The analysis revealed interesting and controversial results showing the some major factors used in creep model predictions [34-37]. The model recommended by the ACI Committee 209 [34] does not require the most significant factor affecting creep which is the mean concrete compressive strength at 28 days because this model calculates the creep coefficient rather than the creep compliance, which may introduce problems due to the assumed value of elastic modulus. This model neglects the importance role of the aggregate-cement ratio and the water-cement ratio which are in the top five factors affecting creep of factors. The models proposed in [36] and [37] require the mean concrete compressive strength at 28 days and the volume-surface ratio but also ignore the role of the aggregate-cement ratio and the water-cement ratio. B3 model [35] requires the top factors affecting creep of concrete. It is important to notice the importance of admixtures in concrete on the results of compliance which in the top 5 of the factors affecting creep of concrete. For that, it is required in future works to present model creep for each admixture depending on its type and percentage/content.

5. Conclusions

Screening design was applied using Daniel Plot, Lenth's Plot and Bayesian screening to examine the most significance factors affecting creep of concrete. This analysis considered experimental results from the NU database. The data used is two-level factorial design of ten factors: water-to-cement ratio, aggregate-to-cement ratio, cement content, mean concrete compressive strength, effective thickness, age at loading, relative humidity, sustained stress, type of cement, and presence of admixtures. The top five individual factors include the mean compressive strength, the volume-surface ratio, the aggregate-cement ratio, the water-cement ratio and the presence of admixtures. None of ACI, CEB and Gardner models require the top 5 factors affecting creep of concrete while calculating the compliance. The analysis shows the major effect of the admixture in concrete as a main factor affecting creep. For that, it is required in future works to present model creep for each admixture depending on its type and its percentage content. The results of this analysis can be used to build accurate computational models to predict creep of concrete by considering the major effects identified in the analysis.

6. References

[1] Bazant Z P 1975 Theory of creep and shrinkage in concrete structures: a Précis of Recent Developments *Mechanics Today* vol. 2 ed S. Nemat-Nasser chapter 1 pp 1-93.
[2] CEB 1993 *CEB Bulletin d'Information*, no. 213/214 (London: Thomas Telford).
[3] Bazant Z P and Baweja S 2000 Creep and shrinkage prediction model for analysis and design of concrete structures—model B3 *The Adam Neville Symp.: Creep and Shrinkage—Structural Design Effects* ed A. Al-Manaseer (Michigan: American Concrete Institute, Farmington Hills) pp 1-83.
[4] Bazant Z P 1999 Criteria for rational prediction of creep and shrinkage of concrete *Creep and Shrinkage of Concrete* ed Ulm F J, Prat M, Calgaro J A and Carol I (Paris: Hermes Science Publications) pp 61-89.
[5] Gvozdev A A 1966 Creep of Concrete *Mekhanika Tverdogo Tela* (Moscow) pp 137-152.
[6] Troxell G E, Raphael J M and Davis R E 1958 Long-term creep and shrinkage tests of plain and reinforced concrete *Proc. ASTM* vol 58 pp 1101-20.
[7] Reddy D H and Ramaswamy A 2018 Experimental and numerical modeling of creep in different types of concrete *J. Heliyon* **4**(7).
[8] ACI Committee 209.1R-05 2005 Report on factors affecting shrinkage and creep of hardened concrete *ACI Committee 209* (American Concrete Institute).
[9] Abd El-Fatah S 2017 Creep of concrete using different cement types and superplasticizer for concrete dams *IJECS* **6**(10) pp 22788-96.
[10] Muller H S and Pristl M 1993 Creep and shrinkage of concrete at variable ambient conditions, in creep and shrinkage of concrete *Proc. 5th International RILEM Symp.* (London,).
[11] Nasser K W and Neville A M 1968 Creep of concrete at elevated temperatures *Proc. ACI J.* **62**(12) pp 1567-79.
[12] Arthananari S and Yu C W 1967 Creep of concrete under uniaxial and biaxial stresses at elevated temperatures *Mag. Concrete Res.* **19**(60) pp 149-56.
[13] Freudenthal A M and Roff F 1958 Creep and creep recovery of concrete under high compressive stress *Proc. ACI J.* **54**(66) pp 1111-42.
[14] Attiogbe E K, See H T and Daczko J A 2002 Engineering properties of self-consolidating concrete *Proc. 1st North American Conference on the Design and Use of Self-Consolidating Concrete* pp 331-6.
[15] Hansen T C and Mattock A H 1966 Influence of size and shape of member on the shrinkage and creep of concrete *Proc. ACI J.* **63**(2) pp 267-89.
[16] L'Hermite R L 1959 What do we know about plastic deformation and creep of Concrete? *RILEM Bulletin* 1(Paris) pp 21-5.
[17] Buil M and Acker P 1989 Creep of a silica fume concrete *Cement and Concrete Res.* **15**(3) pp 463-6.
[18] Mazloom M, Ramezanianpour A A and Brooks J J 2004 Effect of silica fume on mechanical properties of high-strength concrete *Cement and Concrete Comp.*. **26**(4) pp 347-57.
[19] Ross A D 1960 Some problems in concrete construction *Mag. Concrete Res.* **12** pp 27-34.

[20] Bamforth P 1980 In situ masurement of the effect of partial portland cement replacement using either fly ash or ground granulated blast-furnace slag on the performance of mass concrete *Proc. the Institution of Civil Engineers* vol 69(3) pp 777-800.
[21] Lohtia R, Nautiyal B and Jain O 1976 Creep of fly ash concrete *ACI J.* **73**(39) pp 469-72.
[22] Ghosh R S and Timusk J 1981 Creep of fly ash concrete *J. Proc.* **78** pp 351-7.
[23] Collepardi M, Collepardi S and Troli R 2007 Properties of SCC and flowing concrete *Proc. The International Conference of Sustainable Construction Materials and Technologies* (Italy) pp 25–31.
[24] Kristiawan S A and Nugroho A P 2017 Creep behaviour of self-compacting concrete incorporating high volume fly ash and its effect on the long-term deflection of reinforced concrete beam", Procedia Eng. **171** pp 715–24.
[25] Alghazali H H and Myers J J 2015 Creep and shrinkage of ecological self consolidating concrete *Int.Conf. on Performance-Based and Life-Cycle Structural Engineering* (School of Civil Engineering, The University of Queensland) pp1619–27.
[26] ACI Committee 209.2R-08 2008 Guide for modeling and calculating creep and shrinkage in hardened Concrete *ACI Manual of concrete Practice* p 44.
[27] ACI Committee 209R-92 2008 Prediction of creep, shrinkage and temperature effects in Concrete structures *ACI Manual of concrete Practice* p 47.
[28] CEB-FIP Model Code 1990 1993 Model code for concrete structures *Comite Euro-International du Beton (CEB)* (Thomas Telford Services Ltd., London) pp 51~65.
[29] Sakata K 1993 Prediction of concrete creep and shrinkage *Proc. 5th International RILEM Symp. on Creep and Shrinkage of Concrete* (Barcelona, Spain) pp 649-54.
[30] Sakata K, Ayano T and Tanabe 1995 The proposed japanese standard prediction method for creep and shrinkage *ACI Fall Convention* (Montreal, Canada).
[31] Le Roy R, De Larrard F and Pons G 1996 The AFREM code type model for creep and shrinkage of high-performance concrete *Proc. 4th Int. Symp. on Utilization of High-Strength/High-Performance Concrete* vol 2 (Paris) pp 387-96.
[32] Dilger W H, Nitani K and Wang C 1997 A creep and shrinkage model for high-performance concrete *Proc. Int. Conf. on Engineering Materials* vol 1 (Ottawa, Canada) pp 615-30.
[32] Sakata K, Tsubaki T, Inoue S and Ayano T 2002 Prediction equations for creep and drying shrinkage in concrete of wide-ranging strength *Concrete Library International* **40** pp 145-66.
[33] Gilbert R I 2005 AS3600 creep and shrinkage models for normal and high strength concrete, *ACI SP 227-2* pp 21-40.
[34] ACI Committee 209 1992 Prediction of creep, shrinkage, and temperature effects in concrete structures (ACI 211.1-92) *American Concrete Institute*, (Farmingtone Hills, MI) p 47.
[35] Bazant Z P and Baweja S 2000 Creep and shrinkage prediction model for analysis and design of concrete structures: Model B3 *The Adam Neville Symp.: Creep and Shrinkage-Structural Design Effects* ed Al-Manaseer (American Concrete Institue, Farmington Hills, MI) pp 1-83.
[36] CEB 1999 Structural concrete- Textbook on behaviour, design and performance. Updated knowledge of the CEB/FIP model code 1990 *fib Bulletin 2* vol 2 (Federation Internationale du Beton, Lausanne, Switzerland) pp 37-52.
[37] Gardmer N J 2000 Design provisions for shrinkage and creep of concrete *The Adam Neville Symp.: Creep and Shrinkage-Structural Design Effects* ed Al-Manaseer (American Concrete Institue, Farmington Hills, MI) pp 101-34.
[38] Hubler M H, Wender R and Bazant Z P 2015 Comprehensive database for concrete creep and shrinkage: Analysis and recommendation for testing and recording *ACI Mater. J.* **112**(4) pp 547-58.
[39] CEN EN 1992-1-1 2004 Eurocode 2: design of concrete structures- part 1-1: general rules and rules for buildings.
[40] Daniel C 1976 *Applications of Statistics to Industrial Experimentation* (Wiley Sciences, New Jersey).
[41] Lenth R V 1989 Quick and easy analysis of unreplicated factorials *Technometrics* **31**(4) pp 469-73.

[42] Box E P and Meyer R D 1986 An analysis for unreplicated fractional factorials", *Technometrics* **28**(1) pp 11-8.
[43] Box E P and Meyer R D 1993 Finding the active factors in fractionated screening experiments *J. Qual. Technol.* **25**(2) pp 94-105.

IOP Conf. Series: Materials Science and Engineering **809** (2020) 012017 doi:10.1088/1757-899X/809/1/012017

Effect of Humidity on Concrete Creep Predictions

E H Zgheib[1,2,3] and W E Raphael[1]

[1] Ecole Supérieure d'Ingénieurs de Beyrouth (ESIB), Saint Joseph University CST Mkalles Mar Roukoz, Po Box 11-514 Riad El Solh Beirut, 1107 2050, Lebanon
[2] Ecole Doctorale des Sciences et de Technologie (EDST), Lebanese University, Hadath, Lebanon
[3] Current : Lebanese Canadian University, Aintoura, Lebanon

Corresponding author: elise-zg@hotmail.com; elise.zgheib@net.usj.edu.lb

Abstract. An accurate estimation of creep strains is mandatory to maintain the stability and integrity of concrete structures. Different parameters and conditions affect concrete creep predictions more precisely the relative humidity. To study its effect, a large database coming from international laboratories and research centers was applied to compare the experimental measurements to the theoretical values predicted by Eurocode 2 model. An inaccurate estimation was detected, therefore, an amelioration of the Eurocode 2 model was elaborated according to the relative humidity parameter. These calculations allow to predict accurately concrete creep and to avoid prejudicial consequences in structures due to creep phenomenon.

1. Introduction

The concrete is a metastable material that "lives in time". It has a significant time-dependent behaviour. During its lifetime, concrete is subjected to deformations manifested under sustained external load known as creep strains. These strains have an important effect on the stability and integrity of concrete structures. Thus, an accurate estimation of creep strains is required [1, 2].

Different parameters affect concrete creep more precisely the relative humidity which is the scope of this paper. To study its effect, a large worldwide database of creep tests coming from international laboratories and research centers [3] was applied to compare the experimental measurements to the theoretical values predicted by Eurocode 2 (EC2) model using the CEB statistical methods. An inaccurate estimation of the Eurocode 2 creep compliance according to the relative humidity categories is noted.

To overcome this difference, a calibration is required by implementing corrective coefficients to the Eurocode 2 equations. These correction coefficients are quantified by applying the Approximate Bayesian Computation (ABC) method. Using these correction coefficients will help to predict accurately the creep deformations in the design stage and hence, the long-term deflection.

2. Eurocode 2 creep model evaluation

2.1. Eurocode 2 creep model

According to Eurocode 2 model [4], the compliance function can be predicted using the following equation:

$$J(t,t_0)=\frac{1}{E_{cm}(t_0)}+\frac{\phi(t,t_0)}{E_{cm28}} \tag{1}$$

Where; J (t, t_0) is the compliance function, E_{cm28} and $E_{cm}(t_0)$ are the modulus of elasticity of concrete at 28 days after casting and at the loading age t_0 respectively, and φ (t, t_0) is the creep coefficient.

2.2. Creep database

Bažant and Panula were the first researchers that have collected a large worldwide creep and shrinkage database at the Northwestern University (NU) [5]. This database established from American and European institutions consisted of approximately 400 creep tests and 300 shrinkage tests and was included in the papers presenting the BP Model [5]. During the international ACI Fall 1979 Convention, a joint ACI-RILEM committee was organized to extend the Northwestern University database. The pursuit of this work led to the RILEM-ACI 209 database in 1992. After that, in 2008 and 2010, Bažant and Li [6] and Kim [7] have expanded the database at the Northwestern University (NU). The latest database was assembled at NU during 2010-2013 [3]. Information was extracted from many reports, conference proceedings and journal articles. This database is constituted of approximately 1433 creep tests and 1827 shrinkage tests. The tests in the database are performed using different concrete mix composition and placed in various environmental conditions.

2.3. Evaluation methods

The creep compliance J (t, t_0) which is the time-dependent strain per unit stress is calculated according to the Eurocode 2 model for each test in the database. Then the predicted values are compared to the experimental measurements using the CEB mean deviation (M_{CEB}), the CEB coefficient of variation (V_{CEB}), the CEB mean square error (F_{CEB}), and the BP coefficient of variation (ϖ_{BP}). In fact, the CEB statistical indicators were suggested by Muller and Hilsdorf in 1990, while the BP coefficient of variation was developed by Bažant and Panula in 1978 [8].

2.3.1. The CEB mean deviation (M_{CEB}). The M_{CEB} method calculates an average gap and indicates if a model overestimates or underestimates systematically experimental measurements [8]. It may be calculated using the following formulas:

$$M_{CEB}=\frac{\sum_{i=1}^{N} Mi}{N}=\frac{\sum_{i=1}^{N}\left(\frac{1}{n}\sum_{j=1}^{n}\frac{CalX_{ij}}{ObsX_{ij}}\right)}{N} \tag{2}$$

Where; Cal X_{ij} refers to the predicted creep compliance at time j of experiment i, Obs X_{ij} refers to the experimental creep compliance at time j of experiment i, n is the total number of measurements at time j of experiment i, N is the total number of experiments, and M_i is the mean deviation of experiment i.
When M_{CEB} coefficient is near 1, the results of the predicted compliance are close to the experimental values and the model predicts accurately experimental measurements. If the M_{CEB} coefficient exceeds 1, this means that the calculated values are bigger than experimental values and the model overestimates experimental measurements. If the M_{CEB} coefficient is less than 1, then the model underestimates experimental measurements.

2.3.2. The CEB coefficient of variation (V_{CEB}). The V_{CEB} method calculates an average coefficient of variation to evaluate a model relatively to the experimental database. It can be calculated using:

$$Y_i=\frac{\sum_{j=1}^{n} Y_{ij}}{n} \tag{3}$$

$$S_i=\sqrt{\frac{1}{n-1}\times\sum_{j=1}^{n}\left(\Delta Y_{ij}\right)^2} \tag{4}$$

$$V_{CEB} = \sqrt{\frac{1}{N}\left(\sum_{i=1}^{N} V_i^2\right)} = \sqrt{\frac{1}{N}\left(\sum_{i=1}^{N}\left[\frac{S_i}{Y_i}\times 100\right]^2\right)} \quad (5)$$

Small values of V_{CEB} show that the predicted compliance is almost equal to the experimental measurements.

2.3.3. The CEB mean square error (F_{CEB}). The F_{CEB} method calculates the mean square error of the predicted values using the difference between the calculated and observed values relative to the observed value. It may be calculated according to the following formulas:

$$f_j = \frac{(CalX_{ij} - ObsX_{ij})}{(ObsX_{ij})} \times 100 \quad (6)$$

$$F_i = \sqrt{\frac{1}{n-1}\sum_{j=1}^{n} f_j^2} \quad (7)$$

$$F_{CEB} = \sqrt{\frac{1}{N}\sum_{i=1}^{N} F_i^2} \quad (8)$$

Like the coefficient of variation criterion, small values of F_{CEB} show that the predicted compliance is close to the experimental measurements.

2.3.4. The BP coefficient of variation (ϖ_{BP}). This coefficient developed by Bažant and Panula in 1978 is calculated by assigning weight to each point in the data set based on the decade in which it falls and on the number of data points in that particular decade [8]. Data points in each logarithmic decade are considered as one group. The overall coefficient of variation for all data sets is the root mean square of the data set values. The coefficient is given by:

$$\varpi_{ij} = \frac{n}{n_d n_k} \quad (9)$$

$$O_i = \frac{1}{n_w}\sum_{j=1}^{n} \varpi_{ij} \times ObsX_{ij} \quad (10)$$

$$\varpi_i = \frac{1}{O_i}\sqrt{\frac{1}{n-1}\sum_{j=1}^{n} \varpi_{ij}\left(CalX_{ij} - ObsX_{ij}\right)^2} \quad (11)$$

$$\varpi_{BP} = \sqrt{\frac{1}{N}\sum_{i=1}^{N} \varpi_i^2} \quad (12)$$

Where; n_d refers to the number of decades on the logarithmic scale spanned by measured data in data set i, n_k refers to the number of data points in the k-th decade, n_w refers to the sum of the weights of all data points in a data set, ϖ_{ij} refers to the weight assigned to the j-th data point in data set number i, ϖ_i is the coefficient of variation for data set number i, and O_i is the average of the weighted measured value of the creep compliance for the i-th data point in data set number j.
Small values of ϖ_{BP} show that the predicted compliance is close to experimental measurements.

2.4. Results of the Eurocode 2 Evaluation
Since Eurocode 2 model calculates the creep differently according to the concrete strength f_{cm} less than or exceeding 35 MPa and since the relative humidity RH has an important effect on concrete strains [9], Table 1 indicates the evaluation method results for each category as shown below.

Table 1. Eurocode 2 evaluation results.

$\frac{\phi(t,t_0)}{E_{cm28}}$		Evaluation Methods			
		M_{CEB} (expected value 1)	V_{CEB} (expected value 0)	F_{CEB} (expected value 0)	ϖ_{BP} (expected value 0%)
$F_{cm} \leq 35$ MPa	**RH ≤ 60%**	1.52	59.4	128.3	63.60%
	RH > 60%	1.72	148.5	447.4	155.40%
$F_{cm} > 35$ MPa	**RH ≤ 60%**	1.03	52.5	147.8	53%
	RH > 60%	1.24	150.6	213	161.40%

For a concrete with a compressive strength less than 35 MPa, $\frac{\phi(t,t_0)}{E_{cm28}}$ overestimates harmonically the real creep values regardless of the relative humidity. As for compressive strength exceeding 35 MPa, $\frac{\phi(t,t_0)}{E_{cm28}}$ estimates accurately the experimental creep results for a relative humidity less than 60% and overestimates them for RH exceeding 60%.

3. Approximate Bayesian computation (ABC) method

To overcome this inaccurate estimation, correction coefficients are implemented in Eurocode 2 formula as shown in the below equations:

For $f_{cm} \leq 35$ MPa:

$$\frac{\phi(t,t_0)}{E_{cm28}} = O \times \left[\frac{16.8 \times 10^{0.3}}{22000 \times (0.1 + t_0^{0.2})} \times \left[\left(1 + \frac{10}{\sqrt[3]{h_0}} \right) \times \frac{(t-t_0)^{0.3}}{[\beta_h + (t-t_0)]^{0.3}} - \frac{RH \times (t-t_0)^{0.3}}{10 \times \sqrt[3]{h_0} \times [\beta_h + (t-t_0)]^{0.3}} \right] \right] \times f_{cm}^{-0.8 \times P} \quad (13)$$

For $f_{cm} > 35$ MPa:

$$\frac{\phi(t,t_0)}{E_{cm28}} = Q \times \left[\frac{35^{0.2} \times 16.8 \times 10^{0.3}}{22000 \times (0.1 + t_0^{0.2})} \times \left[\left(1 + \frac{(100-RH) \times 35^{0.7}}{(10 \times \sqrt[3]{h_0}) \times f_{cm}^{0.7}} \right) \times \frac{(t-t_0)^{0.3}}{\left[\left(1.5 \times \left[1 + (0.012 \times RH)^{18} \right] \times h_0 + 250 \times \left(\frac{35}{f_{cm}} \right)^{0.5} \right) + (t-t_0) \right]^{0.3}} \right] \right] \times f_{cm}^{-1 \times R} \quad (14)$$

Where O, P, Q and R, are correction coefficients related to each category of f_{cm}.

To quantify these correction coefficients, the Bayesian calibration is applied more specifically the Approximate Bayesian Computation (ABC) method.

The Bayesian calibration provides an automated process for calibrating models by multiplying the expert knowledge known as *a priori* distribution by the likelihood coming from the database [10-12]. Therefore, an *a posteriori* distribution will be defined, which is an update of the knowledge already known by using the latest database. Different methods of Bayesian calibration may be used. In this study, the Approximate Bayesian Computation (ABC) rejection algorithm is applied [13]. The Approximate Bayesian Computation algorithms can be seen as calibration techniques for implicit stochastic models, inferring parameter values in light of computer model, data, prior beliefs about the parameter values and any measurement or model errors [14]. ABC algorithms, sometimes called likelihood-free algorithms, enable inference using only simulations generated from the model, and don't require any evaluation of the likelihood. The most basic form of the ABC algorithm is based on the rejection algorithm. The basic approximate rejection algorithm can be interpreted as performing exact inference in the presence of uniform model or measurement error.

The Approximate Bayesian Computation rejection algorithm is based on generating a random vector for each correction coefficient following an *a priori* distribution [13]. For each random value of the correction coefficient and for all tests in database, the compliance is calculated according to the proposed new equations. These calculations are performed at different time of measurements as provided in the database. Once the updated compliance is calculated, it will be compared to experimental measurement. If the difference between the updated and experimental compliances is less than the threshold, the random variable is accepted; if not, it is rejected. In this study, the rejection parameter is equal to the ratio between predicted and experimental creep. The mean value of the rejection rate is equal to one with a tolerance margin since the predicted values cannot be exactly equal to the experimental measurements. Once this procedure is applied, a set of correction coefficient values is obtained which follows a known *posterior* distribution or an empirical *posterior* distribution.

4. Results

Table 2 summarizes the correction coefficients values that have resulted from the application of the Approximate Bayesian Computation method to the creep coefficient ratio $\frac{\phi(t,t_0)}{E_{cm28}}$ for both categories of f_{cm} (less than 35 MPa and exceeding 35 MPa) and for both subcategories of RH (less than 60% and exceeding 60%).

Table 2. Correction coefficient values.

$\frac{\phi(t,t_0)}{E_{cm28}}$	Evaluation Methods	
	RH ≤ 60%	**RH > 60%**
f_{cm} ≤ 35 MPa	O = 2.5; P = 1.4	O = 2.44; P = 1.45
f_{cm} > 35 MPa	Q = 1; R = 1	Q = 2.2; R = 1.22

It can be noticed that for $f_{cm} \leq 35$ MPa, the correction coefficients are similar for both categories of the relative humidity (RH). As for $f_{cm} > 35$ MPa, and RH < 60%, the correction coefficients are equal to 1 which confirms that no correction is required for this category. Contrary for RH > 60%, the correction coefficients are equal to 2.2 and 1.22 respectively.

To evaluate the accuracy of these correction coefficients, the evaluation methods described in paragraph 2.3 are applied to compare the updated Eurocode 2 formulas with correction coefficients to the experimental measurements.

Table 3. Eurocode 2 evaluation results before and after the Approximate Bayesian Computation (ABC) method correction.

$\frac{\phi(t,t_0)}{E_{cm28}}$		Evaluation Methods							
		M_{CEB} (expected value 1)		V_{CEB} (expected value 0)		F_{CEB} (expected value 0)		ϖ_{BP} (expected value 0%)	
		Before correction	After correction	Before correction	After correction	Before correction	After correction	Before correction	After correction
F_{cm} ≤ 35 MPa	**RH ≤ 60%**	1.52	1.3	59.4	43.6	128.3	104.6	63.60%	47.40%
	RH > 60%	1.72	1.22	148.5	98.5	447.4	302.7	155.40%	102.80%
F_{cm} > 35 MPa	**RH ≤ 60%**	1.03	1.03	52.5	52.5	147.8	147.8	53%	53%
	RH > 60%	1.24	1.14	150.6	150	213	205	161.40%	143.70%

Table 3 summarizes the results of M_{CEB}, V_{CEB}, F_{CEB} and ϖ_{BP} before and after applying the correction to the Eurocode 2 model for each category of the concrete strength and relative humidity using the Approximate Bayesian Computation (ABC) method. It can be noticed that all the above parameters have decreased towards their target values after correction.

IOP Conf. Series: Materials Science and Engineering **809** (2020) 012017 doi:10.1088/1757-899X/809/1/012017

5. Conclusions

Creep strains may affect the stability and integrity of concrete structures if they are not estimated accurately at design stage. To study this phenomenon, a comparison between the compliance predicted by Eurocode 2 and the experimental measurements is performed according to the categories of concrete strength f_{cm} and relative humidity RH using the CEB and ϖ_{BP} evaluation methods. This study shows that Eurocode 2 model overestimates creep in most cases and a correction is required to improve the estimation of concrete creep. Therefore, the Approximate Bayesian Computation (ABC) method is applied to quantify the correction coefficients. The evaluation of the updated model shows an improvement in the accuracy of creep predictions. The Approximate Bayesian Computation rejection algorithm has proven to be an effective solution for the improvement of the creep prediction according to Eurocode 2.

Acknowledgments

The authors greatly acknowledge the National Council for Scientific Research, Lebanon (CNRS-L) and the Research Council of Saint-Joseph University of Beirut for having supported financially this work.

6. References

[1] Su L, Wang Y, Mei S, and Li P 2017 "Experimental investigation on the fundamental behavior of concrete creep" *Constr. and Build. Mater.* **152**, pp 250–8

[2] Zgheib E, Raphael W, and Faddoul R 2018 "Linear Regression Analysis of the Eurocode 2 Creep Model with Bayesian Update" *40th IABSE Symp. 2018* (Nantes, France)

[3] Hubler M H, Wendner R, and Bažant Z P 2015 "Comprehensive database for concrete creep and shrinkage: Analysis and recommendations for testing and recording" *ACI Mater. J.* **112**, pp 547-558

[4] European Committee for Standardization 2004 *Eurocode 2: Design of Concrete Structures – Part 1-1: General Rules and Rules for Buildings* (Brussels, Belgium)

[5] Bažant Z, and Panula L 1978 "Practical prediction of time dependent deformations of concrete, *Matér. et Constr.* **11**, pp 307–28,415–34

[6] Bažant Z, and Li G 2008 "Comprehensive Database on concrete creep and shrinkage," *ACI Mater. J.* **105**, pp 635–8

[7] Kim K 2010 Creep-database V1.1. Available: www.iti.northwestern.edu/publications/bazant/ , https://doi.org/10.21985/N2QN2B

[8] ACI Committee 209 2008 *ACI 209.2R-08 Guide for Modeling and Calculating Shrinkage and Creep in Hardened Concrete* (Farmington Hills, MI, USA: American Concrete Institute)

[9] Raphael W, Faddoul R, Geara F, and Chateauneuf A 2012 "Improvement of the Eurocode 2 Shrinkage Model for Concrete Using a Large Experimental Database." *J. of Struct. Concr. – FIB* **13**, pp 174–181

[10] Box G E P, and Tiao G C 1992, *Bayesian inference in Statistical Analysis* (John Wiley & Sons)

[11] Riddle M, and Muehleisen R 2016 "A guide to Bayesian Calibration of building energy models," *ASHRAE/IBPSA-USA Build. Simul. Conf.* (Atlanta, GA, USA)

[12] Zgheib E, Raphael W, Matar P, Moukarzel I, and Faddoul R 2017 "Approximate Bayesian Computation Method: Update of the Eurocode 2 Creep model. Deterministic and Probabilistic – Rejection Algorithm Approach" *Inter. J. of Civil and Struct. Eng.* **4**, pp 29-33

[13] Robillart M 2015 "*Etude de stratégies de gestion en temps réel pour des bâtiments énergétiquement performants* " Thèse de doctorat de l'Ecole Nationale Supérieure des Mines de Paris, Spécialité Energétique

[14] Wilkinson R D 2013 "Approximate Bayesian computation (ABC) gives exact results under the assumption of model error." *Stat. Appl. in Genet. and Mol. Biol.*, **12**, pp 129–141

IOP Conf. Series: Materials Science and Engineering **809** (2020) 012018 doi:10.1088/1757-899X/809/1/012018

Applicability of Using Natural Fibers for Reinforcing Concrete

Wegdan W. El-Nadoury

Pharos University in Alexandria, Egypt

Wegdan.wagdy@pua.edu.eg

Abstract. The demand for applying sustainability in the construction industry increases the necessity of using natural materials in concrete production. Plant-based natural fibers are cost effective renewable materials which considered a potential way to support sustainable development in both urban and rural regions. This paper investigates the potential of using sugarcane bagasse, palm trunk and banana fibers as concrete reinforcement by determining the effect of different volume percentages of these fiber on the mechanical properties of concrete. The percentage of fiber used was 0.5%, 1.0% and 1.5% for sugarcane, 1%, 2%, and 3% for Palm Trunk, and 0.3%, 0.5%, 0.7%, and 1% for banana. Results showed that the optimum value of sugarcane, and banana is 0.5%, 4% respectively. In addition, the addition of 1%, 2%, and 3% of palm trunk fiber improved the mechanical properties of concrete.

1. Introduction

Concrete has low tensile strength, accordingly, much attention has been given to enhance the mechanical properties of concrete [1]. One of the methods of improving the properties of the concrete is adding fibers [2]. Initial applications incorporate addition of straw to mud bricks, and horse hair to reinforce plaster [3]. The expansion of fiber reinforced concrete began in the early sixties. Currently, growing demand of concrete leads to originating alternate fibers that are sustainable and improve agricultural waste management [4].

Natural fibers like jute, coir, bamboo and sisal have been used as reinforcement materials in concrete mixtures for many years, especially in developing countries [1] [5]. The main benefits of reinforcing concrete with natural fibers are related to low cost of obtaining it and to the compatibility of its properties to fragile materials affirms [5]. Furthermore, natural fibers are renewable natural resources, and thus, do not damage the environment.

The quality of natural fiber reinforced concrete depends mainly on their strength which, in turn, is affected by many factors. The volume percentage of fibers and aspect ratio are considered the most significant factors [6-7]. Lee [8] stated that the addition of fibers to concrete is considered an effective method to improve tensile strength, fatigue resistance, toughness and ductility. Although the incorporation of natural fibers did not eliminate crack formation, they lessen the crack propagation. Lee, Cho and Choi [9] demonstrated that reinforcing composites with fibers delayed the fracture. This is attributed to the formation of transfer bridges by the fiber which absorb part of the stress and result in a more distributed cracking, the rupture becomes a progressive process [10].

The usage of natural fibers for reinforcing the concrete has evidenced valuable improvement in many aspects, nevertheless, it is required to strengthen this study. The aim of this research is to assess the effect of the addition of banana, palm trunk and sugarcane fibers on the mechanical properties of concrete through detecting the compression strength, tensile strength.

2. Experimental Work

2.1 Materials

2.1.1 Binders and Aggregates. The used constituents are ordinary Portland cement, fine aggregate, coarse aggregate and fibers. Ordinary Portland cement (ASTM Type I) is used. Physical properties, chemical analysis, and calculated compound compositions of the used cement are presented in Table 1. The coarse aggregate used in this study is a 19 mm crushed limestone, while the fine aggregate used is siliceous natural sand. The physical characteristics of both aggregates are outlined in Table 2. Both fine and coarse aggregate satisfy the sieved analysis grading in accordance with ASTM C136/C136M-14. The ratio of cement: coarse aggregate: sand is 1:3:1.5. Water-cement ratio was kept constant at 0.5. The mix design of concrete for 1 m^3 is shown in table 3. Table 4 presents the composition of banana, palm trunk, and sugarcane bagasse.

Table 1 The chemical analysis and compound composition of cement

Description	Amount
Setting Time	
Initial:	200 min.
Final:	270 min.
Fineness of cement	3320 (cm^2/gm)
Chemical analysis	
Silicon Dioxide(SiO_2)	20.9
Aluminum Oxide(Al_2O_3)	5.4
Ferric Oxide (Fe_2O_3)	3.4
Calcium Oxide (CaO)	61.28
Magnesium Oxide (MgO)	2.6
Sulfur Trioxide (SO_3)	1.8
Insoluble residue	0.62
Loss on ignition	1.8
Calculated Compound Composition	55.9
Tricalcium Silicate (C_3S)	19
Dicalcium Silicate (C_2S)	7.5
Tricalcium Aluminate (C_3A)	9.8
Tetracalcium aluminoterrite (C_4AF)	

Table 2. The physical characteristics both aggregate

Property	Sand	Coarse Aggregate
Unit Weight(t/m^3)	1.77	1.5
Specific weight	2.62	2.6
Fineness Modulus	2.65	…..
Nominal Maximum Aggregate Size	………	19 mm

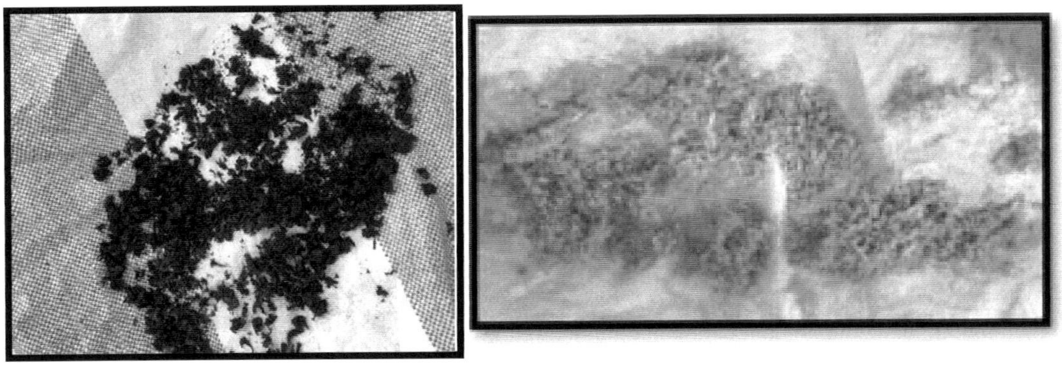

(a) (b)

Figure 1. Fibers after drying in the oven for 24 hours (a) Banana fibers (b) Sugarcane bagasse.

Table 3. Mix design of concrete for $1m^3$

Water(kg/m^3	Cement(kg/m^3)	Sand(kg/m^3)	Coarse aggregate(kg/m^3)
225	450	675	1350

2.1.2 Fibers. The fibers used in this work are banana, palm trunk and sugarcane bagasse. Before being incorporated into the concrete the fibers are treated. As for Banana and sugarcane bagasse fibers, they are subjected to chemical treatment in order to increase the pH value of the fibers. Alkali treatment of cellulosic fibers is the usual method to produce high quality fibers. The sugarcane bagasse will dry under the sun until completely dry for 7 days. Then the bagasse will cut into small relatively uniform strips with estimation 5cm until 10cm. Both banana and sugarcane bagasse fibers are cleaned and submersed sodium hydroxide solution for two hours at room temperature as shown in Figure. 1. The concentration of sodium hydroxide solution is 6% and 50% for banana and sugarcane respectively. The fibers are then immersed in water to be washed thoroughly to remove the unreacted alkali and then dried in an oven at 80 °C for 24 h. Concerning the palm trunk fibers, they are taken directly from the plantation when the trees are recently fallen. The fibers with length vary from 25-300 mm are processed manually. The tested percentages for sugarcane are; 0.5%, 1%, and 1.5%, while for palm trunk are; 1%, 2%, and 3%, finally percentages for banana are; 0.3%, 0.5%, 0.7%, and 1%.

Figure 2. Specimens for compression and tensile strength test.

3. Specimens Preparation

Mixing concrete is performed by using a free-fall mixer, the natural fibers are the final material to be added in the mixture. Subsequently, 50 specimens are produced in the dimensions of 15 cm x 15 cm x 15 cm, as well as 15 prismatic specimens with a section of 4 cm x 4 cm and a length of 16 cm, as shown in Figure 2. The specimens are demolded after 24 hours and then cured in water for 28 days. The mechanical properties evaluations of the reinforced concrete with natural fibers are obtained through compressive strength, and tensile strength tests.

4. Test Procedures

4.1 Compressive strength

The axial compression test consists of determining the maximum load of rupture supported by the specimen. For the accomplishment of this mechanical test, 3 test specimens of each type of concrete used in this study are tested.

4.2 Tensile strength

In order to perform this test, a prismatic section specimen is submitted to bending, with loading in two symmetrical sections, until rupture. A total of 3 specimens of each concrete type are tested.

5. Results

5.1 Compressive strength

The control sample containing 0% fiber records a reading of 30 MPa. For concrete with sugarcane bagasse, samples containing 0.5%, 1.0%, and 1.5% fiber have compressive strength of 30.30, 31, 32 MPa respectively. For concrete with palm trunk, samples containing 1%, 2%, and 3% fiber have compressive strength of 40, 26, 21 MPa respectively. For concrete with banana fibers, samples containing 0.3%, 0.5%, 0.7%, and 1%, the compressive strengths are; 30.8, 38.7, 35, and 33 respectively. The effect of sugarcane bagasse, Palm trunk, and banana fibers are illustrated in Figure 3.

It can be seen that the addition of sugarcane bagasse has a minor effect on concrete compressive strength. However, the increase in compressive strength is directly proportional to the percentage of fiber added. This is I contrast to the results for concrete with banana fibers and palm trunk fiber. In both cases, the concrete compressive strength is enhanced till optimum percentage of fibers and then dimensioned. This may be attributed to the weak interfacial bond between fiber and concrete due to the formation of voids. From the previously mentioned discussion, it is obvious that the optimum percentages of banana fibers and palm trunk fiber are; 0.5% and 1% respectively. While more experiments should be done to assess the effect of higher percentages of sugarcane fibers on compressive strength.

Table 4. The composition of Palm trunk, sugarcane bagasse, and banana fiber

Fiber	Palm Trunk	Sugarcane bagasse	Banana
Component	Percentage	Percentage	percentage
Cellulose	50 %	66	60
Lignin	25%	21	15
Extractives	12%	6	9
Moisture	8%	----	10
Ashes	5%	7	6

IOP Conf. Series: Materials Science and Engineering **809** (2020) 012018 doi:10.1088/1757-899X/809/1/012018

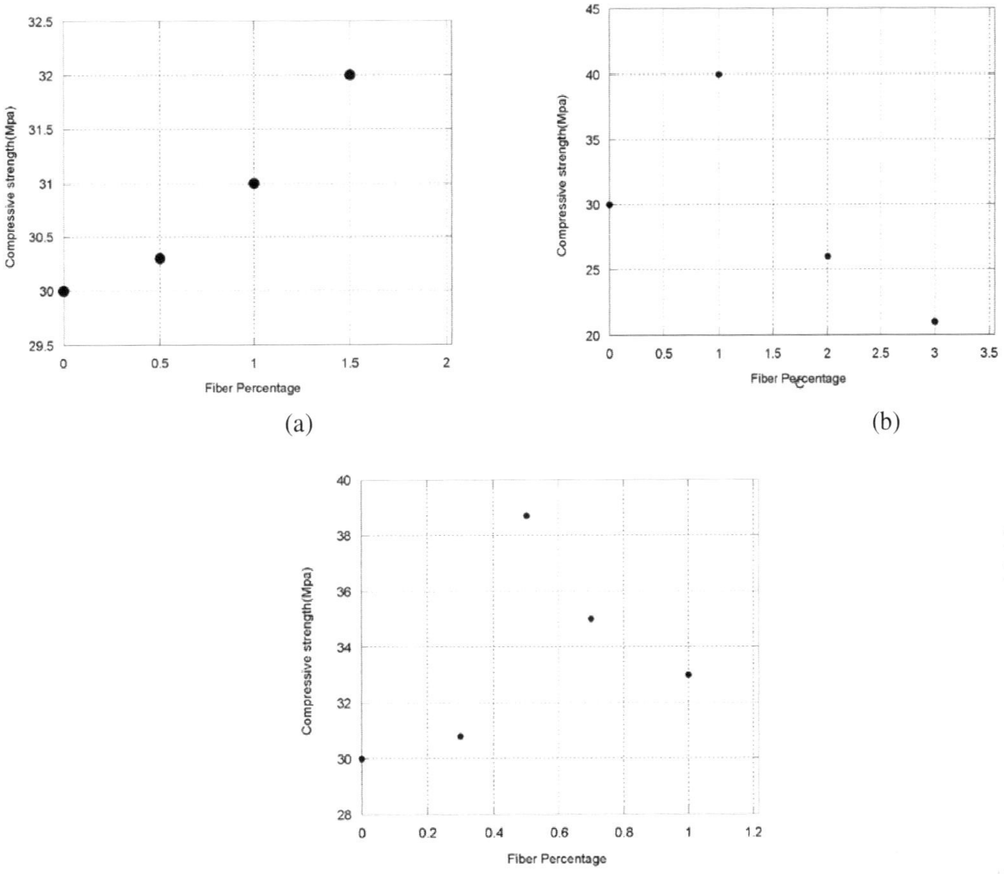

(a) (b)

(c)

Figure 3. Effect of fiber on compressive strength (a) sugarcane bagasse, (b)Palm trunk, (c) banana fibers

5.2 Tensile strength

The control sample of normal concrete containing 0% fiber records a reading of 2.10 MPa tensile strength. For concrete with sugarcane bagasse, samples containing 0.5%, 1.0%, and 1.5% fiber has tensile strength of 2.15, 2.25, 2.3 MPa respectively. For concrete with palm trunk, samples containing 1%, 2%, and 3% fiber has tensile strength of 2.5, 2.08, 2.02 MPa respectively. For concrete with banana fibers, samples containing 0.3%, 0.5%, 0.7%, and 1%, record tensile strength of 2.4, 2.54, 2.72, and 2.9 respectively. As it can be seen from Figure 4, the tensile strength is directly proportional to the increase in percentages of fiber added either for banana fibers or for sugarcane bagasse. The addition of banana fibers has the highest effect on tensile strength. As for concrete reinforced with palm trunk fibers, the optimum percentage for fiber is 1%. This may be attributed to difficulty in compacting concrete containing a high percentage of palm trunk, accordingly, more voids and less bond between fiber and concrete is achieved. However, all concrete samples reinforced with the three types of fibers do not face complete rupture even under the ultimate load the maximum load was applied, the fibers diminish the crack mouth opening.

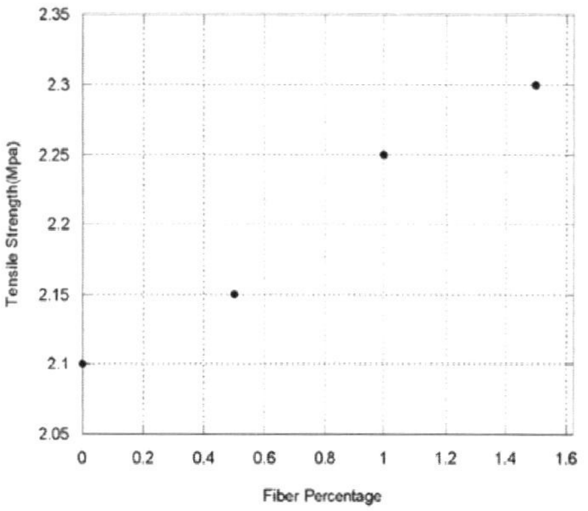

(a)

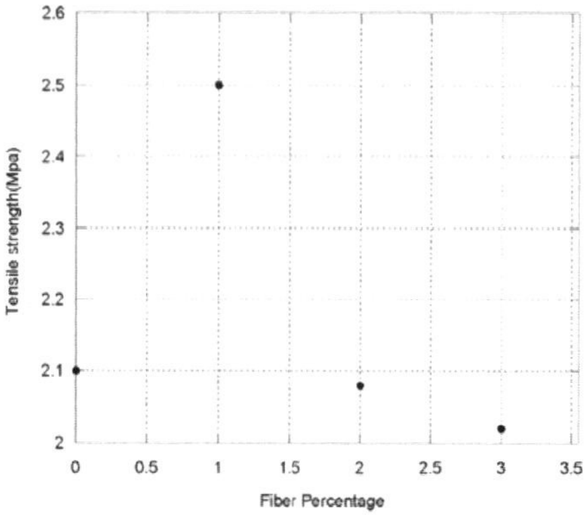

(b)

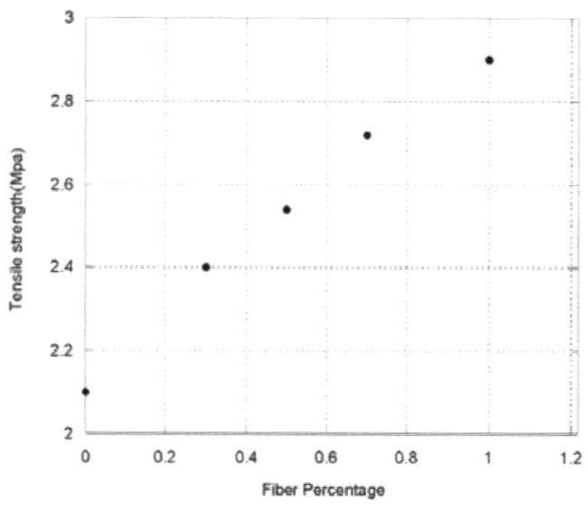

(c)

Figure 4 Effect of fiber on tensile strength (a) sugarcane bagasse, (b) Palm trunk, (c) banana fibers

6. Conclusion

The addition of natural fibers improves the mechanical properties of concrete.

Addition of Banana fiber with 0.5% increase the compressive strength and tensile strength by 30% and 14% respectively compared to the control mixture. While the addition of 1% increases the tensile strength by 38% compared to control mixtures.

Addition of sugarcane bagasse increases both compressive and tensile strength but to lower limit compared to the other two types of fibers. The maximum increase in compressive and tensile strength is 7%, 9.5%.

Compressive strength of the concrete decreases when the amount of palm trunk fiber increased. The optimum percentage that gives the higher compressive strength is 1%. The same trend is observed in testing the tensile strength. The addition of this percentage increases the compressive strength by 33% and the tensile strength by 19% compared to the control mixture.

Acknowledgment
The authors wish to thank Pharos University; the experimental work has been done in laboratories of the university and using the resources of the university.

7. References

1. S. Chen, B. Mulgrew, and P. M. Grant, "A clustering technique Balaguru, P.N., Shah, S.P., 1995. Fiber reinforced cement composites. McGraw-Hill.
2. Oraimi SK, Seibi AC. Mechanical characterization and impact Ramakrishna, G and Sundarajan, T., Impact strength of a few natural fibre reinforced cement mortar slabs: a comparative study, Cem. Concr. Compos. 2005; 27(5):547-553.
3. Savastano Jr H, Agopyan V, Adrian MN, pimental L. Plant fiber-reinforced cement components for roofing, Constr Build Mater. 1999; 13:433-438.
4. Savastano Jr H, Warden PG, Coutts RSP. Microstructure and mechanical properties of waste fibre-cement composites, Cem. Concr. Compos. 2005; 27(5):583-592.
5. Wolfe, R.W., Gjinolli, A., 1996. Cement bonded wood composites as an engineering material. In: The use of recycled wood and paper in building applications. Madison, WI: Forest Products Society, 84–91.
6. Ramirez-Coretti, A., 1992. Physical-mechanical properties of fiber cement elements made of rice straw, sugar cane bagasse, bananaracquis and coconut husk fibers, Fiber Reinforced Cement and Concrete, Ed: Swamy R.N. E&FN Spon.
7. Jorillo, P., Shimizu, G., 1992. Coir fiber reinforced cement based composite, Part2: fresh and mechanical properties of fiber concrete, Fiber Reinforced Cement and Concrete, Ed: Swamy, E&FN Spon.
8. Lee JH. Influence of concrete strength combined with fiber content in the residual flexural strengths of fiber reinforced concrete. Composite Structures. 2017;168:216-225.
9. ulien Bras, Mohammad L. Hassan, Cecile Bruzesse, Enas A. Hassan, Nahla A. El-Wakil, Alain Dufresne. Mechanical, barrier, and biodegradability properties of bagasse cellulose whiskers reinforced natural rubber nanocomposites, Industrial Crops and Products 32 2010, 627 – 633
10. Lee JH, Cho B, Choi E. Flexural capacity of fiber reinforced concrete with a consideration of concrete strength and fiber content. Construction and Building Materials. 2017;138:222-23

IOP Conf. Series: Materials Science and Engineering **809** (2020) 012019 doi:10.1088/1757-899X/809/1/012019

Preliminary Study on the Treatment of Benzene Contaminated Water using an Argon Microwave Plasma Jet

Enrique Casado,[1] María C. García,[2] Dorota A. Krawczyk[3], and Antonio Rodero[1]

[1]Department of Physics, Universidad de Córdoba,Córdoba, Spain.
[2]Department of Applied Physics, Universidad de Córdoba, Córdoba, Spain.
[3]Bialystok University of Technology, Wiejska 45 E, 15-351 Bialystok, Poland

fa1rosea@uco.es

Abstract. In this work, the ability of a microwave induced plasma jet to treat water contaminated with benzene was investigated. It studies the formation of hydroxyl (OH) radicals by the plasma jet and reactions between benzene and these radicals, under different experimental conditions (including microwave power and treatment time). Products resulting from the interaction plasma-sample were identified and quantified using Gas Chromatography – Mass Spectrometry (CG-MS) analysis.

1. Introduction

In the last years, the use of plasma technology to induce Advances Oxidation Processes for water treatment purposes has notably grown [1-3]. Plasmas generate charged particles, excited species, radicals and UV radiation, among others, each of which are advanced oxidation techniques themselves. These species are able to degrade contaminants through mineralization [4-8]. In this work we report on the degradation of benzene by an argon microwave induced plasma jet.

2. Materials and method

2.1. Plasma generation

An argon plasma was created using a *surfatron* microwave power coupling device [9]. A T-shaped quartz tube (1.5 and 4 mm of inner and outer diameter) with one of its ends opened to the air was used (see figure 1). This configuration forced the plasma column to go down through the vertical part of the T-tube making it easier to control the plasma-liquid distance.

2.2. Liquid samples.

Liquid samples suffered a remote plasma exposure and plasma-sample distance was fixed at 2 cm. The argon flow rate was set at 1.5 L/min, and the microwave power was changed from 30 to 60 W. Different treatment times of the sample were also explored. Liquid samples were placed in cylinder shaped glass containers (2 cm diameter and 4 cm high). In OH generation studies, 4 ml of miliQ water with 3-Coumarin Carboxylic acid dissolved was used. The concentration of the solution was 0.003 M. In the second part of this work, 1.5 ml of benzene was added to the miliQ water, staying on the top.

IOP Conf. Series: Materials Science and Engineering **809** (2020) 012019 doi:10.1088/1757-899X/809/1/012019

2.3. Analysis of samples.

OH radicals generation rate was measured from the fluorometric method proposed by Newton and Milligan [3] based on the use of 3-Coumarin Carboxylic acid dissolved in water. This component reacts with OH radicals, being 7-Hydroxycoumarin Carboxylic acid (which is fluorescent) one of the resulting products. Thus, the method consists of treating water samples with 3-Coumarin Carboxylic acid dissolved and see how much 7-Hydroxycoumarin Carboxylic acid is produced, using a spectrofluorometer. From the rate formation of this compound and taking into account that the yield per OH radical of this fluorescent product is about 4.7%, the rate formation of OH radicals can be measured [10]. A PTI, Quanta Master 40 UC/VIS Steady State Spectrofluorometer was used to measure the concentration of the fluorescent product.

Solutions of benzene treated with this plasma reactor were analysed using a GC-MS (Thermo Finnigan, Thermo-Quest Trace GC/MS–Trace DSQ) and changes of product concentrations in this solution were obtained. Evaporation produced in the sample due to high volatility of toluene was taken into account in terms of changes in concentration.

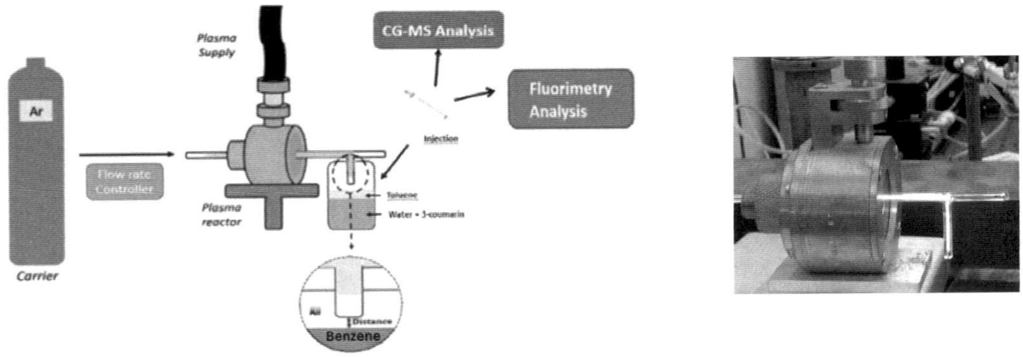

Figure 1. Experimental set-up.

3. Results

3.1. OH radical generation

Solutions with 3-Coumarin Caboxylic acid were treated with the plasma reactor under different experimental conditions, and formation of 7-Hydroxycoumarin Carboxylic acid was studied. Changes in concentration of this compound were analysed.

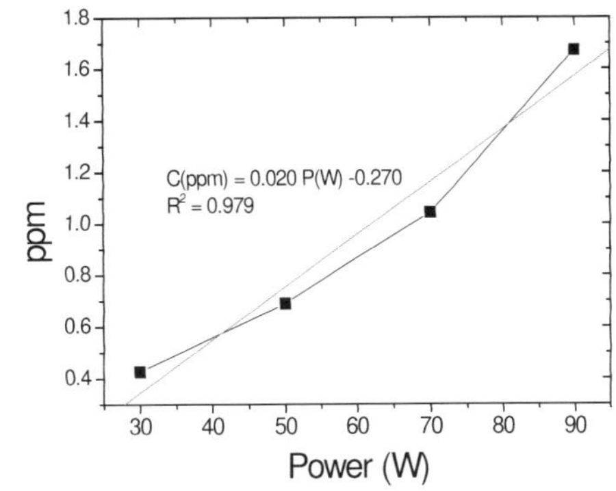

Figure 2. Concentration in ppm of 7-Hydroxycoumarin Carboxylic acid for different powers.

In first instance, the dependency on the microwave power injected to the plasma was studied (see figure 2). Microwave power was set at 30, 50, 70 and 90 W levels, while treatment time, gas flow rate and distance were kept constant. A linear relationship was found between 7-Hydroxycoumarin Carboxylic acid concentration and power with a rate of 0.02 ppm/W, which could be ascribed to a higher OH production by plasma for higher microwave power.

Figure 3 depicts changes in 7-Hydroxycoumarin Carboxylic acid concentration of the sample after different treatment times (1, 1.5, 2 and 2.5 min), for a distance, microwave power and gas flow rate 1.5 L/min fixed. In this case, a linear dependency was also found with a rate of 0.61 ppm/min.

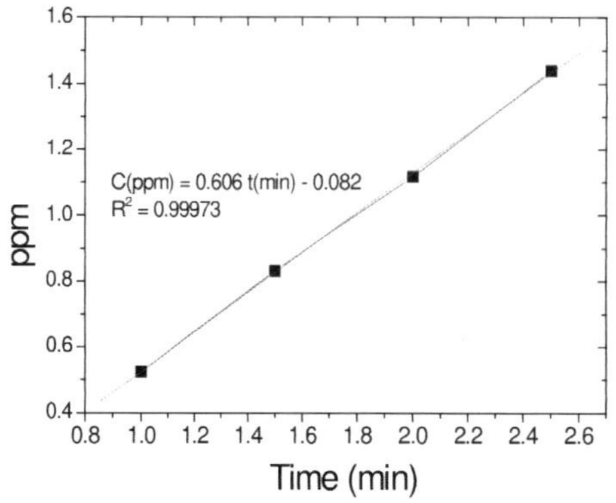

Figure 3. Concentration in ppm of 7-Hydroxycoumarin Carboxylic acid for different treatment times.

3.2. Oxidation of Benzene

Products resulting from the interaction plasma-benzene were identified and quantified using CG-MS analysis. In the present case, the attention was focused on two products resulting from oxidation of benzene, phenol and 1,2-benzenediol, whose peaks appeared at 3.31 min and 5.76 min, respectively.

Phenol and 1,2-benzenediol came from benzene oxidation. Formation of these products under different experimental conditions of microwave power and treatment time was also studied. The areas of their corresponding GC-MS peaks were measured for this purpose.

Figure 4 shows changes in phenol and 1,2-benzenediol content of the treated samples under different microwave power conditions (from 30 to 60 W). An increasing trend was found according to the dependence on the microwave power of OH radical formation measured (Figure 2).

Figure 5 shows changes of phenol and 1,2-benzenediol content after different times of treatment (0.5, 1, 1.5 and 2 min), for a plasma-sample distance, microwave power and gas flow rate 1.5 L/min fixed. A linear dependency was found, which agrees with results shown in figure 3 for 3-Coumarin Carboxylic acid oxidation.

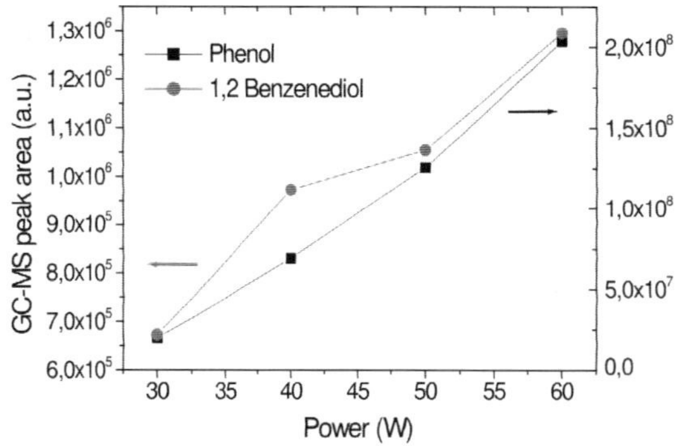

Figure 4. Phenol and 1,2-benzenediol GC-MS peak area for different microwave powers (F = 1.5 L/min, t = 1 min, d = 2 cm).

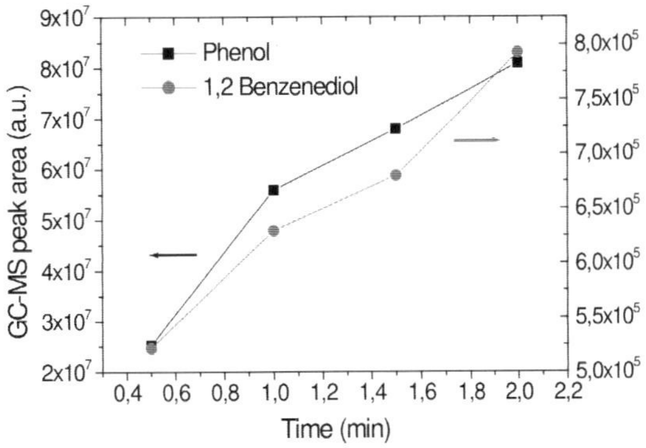

Figure 5. Phenol and 1,2-benzenediol GC-MS concentration for different treatment times (P = 50 W, F = 1.5 L/min, d = 2 cm).

4. Conclusions

This work proves the generation of OH radicals when an argon microwave plasma jet is being used for water treatment, in a remote exposure. The amount of OH radicals generated depends upon plasma microwave power and treatment time.

Interaction plasma-benzene cause oxidation of benzene, being phenol and 1,2-benzendiol the main products identified. This oxidation is controlled by OH production. These preliminary results show that the microwave plasma jet is a promising technology for treatment of benzene contaminated water.

Acknowledgements

As part Authors thank the European Regional Development Funds program (EU-FEDER) and the MINECO (project MAT2016-79866-R) for financial support. The authors also are grateful to Andalusian Regional Government (Research Group FQM-136) for their technical and financial support. This work was also partially financed from the project BUT InterAcademic Partnerships (PPI/APM/2018/1/00033/DEC/1).

5. References

[1] Kornev I and Preis S 2016 *J. Adv. Oxid. Technol.* **19** 284-289

[2] Rodriguez-Narvaez OM, Peralta-Hernandez JM, Goonetilleke A and Bandala ER 2017 *Chem. Eng.J.* **323** 361–380

[3] Yang Y, Ok YS, Kim KH, Kwon EE and Tsang YF 2017 *Sci. Total Environ.* **596–597** 303–320

[4] Foster J, Sommers BS, Gucker SN, Blankson IM and Adamovsky G 2012 *IEEE Trans. Plasm. Sci.* **40** 1311-13123

[5] Ceriani E, Marotta E, Shapoval V, Favaro G and Paradisi C 2018 *Chem. Engin. J.* **337** 567–575

[6] Tampieri F, Giardina A, Bosi FJ, Pavanello A, Marotta E, Zaniol B, Neretti G and Paradisi C 2018 *Plasma Process. Polym.* 2018 15:e1700207

[7] Magureanu M, Mandache NB and Parvulescu VI 2015 *Water Res.* **81** 124–136

[8] Garcia MC, Mora M, Esquivel D, Foster JE, Rodero A, Jiménez-Sanchidrián C, Romero-Salguero FJ 2017 Chemosphere **180** 239-246

[9] Moisan M and Pelletier J 1992 *Microwave Excited Plasmas* (Amsterdam: Elsevier) Chap. 3

[10] Newton GL and Milligan JR 2006 *Radiation Physics and Chemistry* **75** 473-478

Obtaining Chitin/Chitosan-Melanin Complexes from Black Soldier Fly *Hermetia Illucens*

Adelya Khayrova[1,2,a], **Sergei Lopatin**[1,b] **and Valery Varlamov**[1,c]

[1]Institute of Bioengineering, Research Centre of Biotechnology of the Russian Academy of Sciences, Leninsky prospect, 33, build. 2, 119071, Moscow, Russia
[2]Entoprotech LLC, Nobel street, 5, Skolkovo innovation centre, Moscow, Russia

[a]adelya15@mail.ru, [b]lopatin@biengi.ac.ru, [c]varlamov@biengi.ac.ru

Abstract. Chitosan, obtained by deacetylation reaction of chitin, is a regular biopolymer with unique properties, including high biological and sorption activities. Melanin is considered as a natural photo and radio protector. Depending on the source and the isolation method, melanin possesses various physicochemical properties that can complement the radioprotective activity of chitosan and be used to solve important scientific and practical issues. Insects are relatively understudied natural sources of chitin-melanin complexes, whereas these components are linked by strong covalent bonds. This study developed a technology for the production of this natural conjugate, based on which its deacetylated derivative, the chitosan-melanin complex, was obtained, taking the black soldier fly as an example.

1. Introduction

The discovery of new and sustainable approaches for efficiently utilising biomass resources is a global priority, with the world population predicted to reach 9 billion in 2050. Due to its ability to feed on a wide range of organic waste, the black soldier fly *Hermetia illucens* has emerged as the most important insect for bioconversion [1]. Therefore, this insect species is becoming a valuable tool for protein production and waste management, and several companies around the world are interested in exploring its potential [2]. Insects, in addition to being natural recyclers of organic waste, are efficient producers of protein, fat and other important bioactive substances such as chitin and melanin. Other advantages of insects include their positive environmental, economic and ecological impacts. Farming insects has many benefits compared with conventional livestock farming approaches such as increased feed-conversion efficiency, lower greenhouse gas emission (by up to 100 times) and water pollution, and reduced land use with low environmental contamination [3].

Chitin is the second most common polysaccharide in nature after cellulose. The recent emphasis on environmentally friendly technology has led to increased interest in biopolymers and bio-based polymers. They are more beneficial compared to synthetic polymers due to increased functionality and biodegradability. Chitosan can be obtained from chitin by deacetylation reaction, offering a wide range of applications in many fields including biotechnology, food industry, and medicine due to its exceptional properties such as biocompatibility, biodegradability, non-toxicity, and chelation of metal ions [4].

The exoskeleton of insects provides structural support and is made up of cuticle, an extracellular matrix comprising chitin fibers, cuticular proteins, lipids, and pigments [5]. One of the most common pigments found in insect cuticle is melanin, which is crucial for colouration, wound healing and innate immunity [6]. Within the animal kingdom, melanins are classified based on the presence of a group, into eumelanin (which imparts brown to black colouration) and phleomelanin (red to yellow colouration) [7]. It is known that melanins are photo-protective, providing the protection for skin against the harmful effects of ultraviolet (UV) and visible radiation [8]. As a result, its optical properties have been widely investigated for the past years. Being able to dissipate 99% of the absorbed UV and visible radiation through non-radiative means, melanin is considered as an optimal photo-protector [9].
Based on this information, it can be assumed that when complexed with chitosan, melanin would exhibit increased biological activity. In this study, pupal exuviae and dead imagoes of the black soldier fly *Hermetia illucens* were investigated as melanin has already been developed at these stages and covalently bound to chitosan.

2. Materials and methods

Hermetia illucens pupal exuviae and dead imagoes were provided by Entoprotech LLC, Russia. Samples consisting of pupal exuviae and dead imagoes were powdered into 35-70 mesh size.

2.1. Obtaining chitin/chitosan-melanin complexes from pupal exuviae

2.1.1. Demineralisation. 1000 mL of 1% HCl was added to 100 g of pupal exuviae and left at 20 °C for 2 h. The solid residue was separated through glass filter. It was washed with distilled water and lyophilised afterwards. 74 g of product was obtained (yield=74%).
2.1.2. Deproteinisation and defatting. 300 mL of 5% (w/v) KOH was added to 30 g of demineralised biomass and left in 50 °C water bath for 2 h with occasional stirring. The solid residue was separated through glass filter, washed with distilled water until neutral pH and lyophilised. 23 g of product was obtained (yield=77%). Treatment with stronger alkali, 10% (w/v) KOH and 50% (w/v) NaOH, was carried out in the same way. 10 g (yield=34%) and 9 g (yield=30%) of product was obtained, respectively.
2.1.3. Deproteinisation, defatting and deacetylation. Performed by using 80 mL of 50% (w/w) NaOH. The alkaline solution was added to 4 g of chitin-melanin complex and left at room temperature for 30 min. The suspension was warmed in 100 °C water bath for 2 h with occasional stirring. The suspension was cooled, washed until neutral pH and lyophilised. 1.2 g of chitosan-melanin complex was obtained (yield=29%).

2.2. Obtaining chitin/chitosan-melanin complexes from dead imagoes

2.2.1. Demineralisation. 1000 mL of 1% HCl was added to 100 g of dead imagoes and left at 20 °C for 2 h. The solid residue was separated through glass filter. It was washed with distilled water and lyophilised afterwards. 87 g of product was obtained (yield=87%).
2.2.2. Deproteinisation and defatting. 300 mL of 5% (w/v) KOH was added to 30 g of demineralised biomass and left in 50 °C water bath for 2 h with occasional stirring. The solid residue was separated through glass filter, washed with distilled water until neutral pH and lyophilised. 7.2 g of product was obtained (yield=24%). Treatment with stronger alkali, 10% (w/v) KOH and 50% (w/v) NaOH, was carried out in the same way. 4.5 g (yield=15%) and 2.7 g (yield=9%) of product was obtained, respectively.
2.2.3. Deproteinisation, defatting and deacetylation. Performed by using 40 mL of 50% (w/w) NaOH. The alkaline solution was added to 4 g of chitin-melanin complex and left at room temperature for 30 min. The suspension was warmed in 100 °C water bath for 2 h with occasional stirring. The suspension was cooled, washed until neutral pH and lyophilised. 0.72 g of chitosan-melanin complex was obtained (yield=18%).

3. Characterisation

3.1. Ash content determination

The ash content was determined gravimetrically by incinerating the sample (0.6-6 g) in a muffle furnace at 650 °C for 4 h. The value of ash content was found as a proportion of weight of the residue to the weight of the sample according to Equation 1 below:

$$\%ash = (\frac{m_2}{m_1}) \times 100, \quad (1)$$

where m_1 and m_2 are weights of the initial sample and the residue, respectively [10].

3.2. Fat content determination

Fat content was estimated by Soxhlet extraction method. The fat value was measured using Equation 2:

$$\%fat = (\frac{m_2 - m_1}{m_1}) \times 100, \quad (2)$$

where m_1 is the initial mass of the sample, while m_2 – mass of the sample after being dissolved in diethyl ether.

3.3. Amino acid concentration determination

Ion-exchange chromatography coupled with post-column derivatisation of samples with ninhydrin was used to determine the total content of amino acids (free and bound). Acid hydrolysis (with the exception of tryptophan) in 6 M HCl solution was used for sample preparation, with norleucine added as an internal standard. The hydrolysis was carried out at 110 ° C for 24 h. The samples were oxidised with 50% n-formic acid (HCOOOH) to determine cystine and methionine.

Alkaline hydrolysis in $Ba(OH)_2$ solution was used for tryptophan determination, with norleucine added as an internal standard at 110 ° C for 20 h.

The analysis was performed on Shimadzu LC-20 Prominence high performance liquid chromatography (HPLC) system (Japan), with a reaction module for post-column derivatisation with ARM-1000 ninhydrin (Sevko & Co, Russia) equipped with an absorption detector (λ_{abs} = 440 nm, 570 nm) and column with ion exchange resin 4.6 x 150 mm (Sevko & Co, Russia). Ready-to-use buffer solutions (Sevko & Co, Russia) were provided.

The concentration was calculated using a standard amino acid sample (Sykam, Germany).

4. Results and discussion

Pupal exuviae and dead imagoes of *H. illucens* were subjected to demineralisation, deproteinisation, defatting and deacetylation steps. The results of physicochemical analyses are presented in Table 1. It is important to note that the impurities in the chitin/chitosan-melanin complexes have not been previously evaluated in the literature [11].

It was assumed that the conditions (1-10% NaOH), previously used for crustaceans and bees, were suitable for removal of protein and fat from pupal exuviae and dead imagoes [12]. Two concentrations of potassium hydroxide were initially used – 5 and 10% (w/v) – after a demineralisation step in 1% HCl. However, this treatment did not allow to remove the majority of protein and fat residues from the chitin-melanin complexes, therefore, the alkali concentration was increased to 50% (w/v) and (w/w). The protein content is shown in Table 2 and was determined as a total amount of amino acids present in a sample. Dead imagoes contained a particularly high amount of fat (21-27%) after the alkali treatment. This can be explained by the fact that the fat in dead imagoes differs in composition and may contain wax. Hence, an additional defatting step using an organic solvent was introduced after demineralisation. As a result of treatment with diethyl ether, the fat from the chitosan-melanin complex was removed. Previously, we have shown that the treatment of chitin from *H. illucens* with 50% (w/w) alkali allows to obtain chitosan with degree of deacetylation of 90%, whereas treatment with weaker alkali (50% (w/v)) does not lead to deacetylation of chitin [13]. It was observed that as the concentration of NaOH increased, the protein content of the dead imagoes also increased. It could be due to the fact that the more concentrated alkali is denser and more viscous, which does not allow it to effectively wet the material and penetrate into it. Based on these findings, it is preferred to obtain chitosan-melanin complex out of chitin-melanin complex, instead of the demineralised material.

In addition, it was observed that the pupal exuvia has a two-layer structure. The inner white layer is chitin, and the outer dark layer is the chitin-melanin complex. (Figure 1) The inner chitin layer can be removed by direct extraction using phosphoric acid [14].

A mixture of chitosan and chitosan-melanin complex can be obtained by deacetylation reaction. Chitosan can be separated from the chitosan-melanin complex by dissolving it in acetic acid.

Table 1. Physicochemical analyses of pupal exuviae and dead imagoes at various treatment stages.

Treatment	Pupal exuviae			Dead imagoes		
	%fat	%ash	%protein	%fat	%ash	%protein
Starting material	8.91	10.51		27.08	5.16	
Demineralisation (1% HCl)	11.26	1.67		26.63	2.71	
5% (w/v) KOH	0.86		21.78	21.17		19.88
10% (w/v) KOH	5.45		11.57	26.63		11.45
50% (w/v) NaOH	0.40		1.69	7.80		3.82
50% (w/w) NaOH	0		2.83	5.40		8.18
Defatting ($(C_2H_5)_2O$)	-			0		

Table 2. Amino acid profiles of pupal exuviae and dead imagoes after the alkali treatment.

Amino acid	Pupal exuviae		Dead imagoes	
	50% (w/w) NaOH	50% (w/v) NaOH	50% (w/w) NaOH	50% (w/v) NaOH
Aspartic acid (Asp)	0.40	0.47	0.93	0.35
Threonine (Thr)	0.05	0.03	0.14	-
Serine (Ser)	0.06	0.03	0.08	-
Glutamic acid (Glu)	0.15	0.16	0.69	-
Glycine (Gly)	0.10	0.07	0.13	-
Alanine (Ala)	-	0.01	0.07	-
Valine (Val)	0.17	0.12	0.41	0.26
Isoleucine (Ile)	0.05	0.03	0.14	0.14
Leucine (Leu)	0.12	0.08	0.41	0.30
Tyrosine (Tyr)	-	-	-	-
Phenylalanine (Phe)	-	-	0.26	0.17
Histidine (His)	0.49	0.27	0.40	0.55
Lysine (Lys)	-	-	0.42	0.20
Arginine (Arg)	-	-	0.34	-
Proline (Pro)	1.10	0.36	3.68	1.76
Cysteine (Cys)	0.14	0.06	0.08	0.09
Methionine (Met)	-	-	-	-
Total	2.83	1.69	8.18	3.82

Figure 1. The picture of pupal exuviae (on the left), chitin (bottom right) and chitin-melanin complex (upper right).

5. Conclusion

In this work chitin- and chitosan-melanin complexes from *Hermetia illucens* were obtained and characterised for the first time. Optimal conditions for the removal of impurities were adjusted. It is expected that chitin/chitosan-melanin complexes may have higher biological and sorption activities compared to pure chitin and chitosan and are of great interest for future studies.

6. Acknowledgements

The present research was carried out within the framework of the state assignment and partially supported by Entoprotech LLC.

7. References

[1] Müller A, Wolf D and Gutzeit H, "The black soldier fly, Hermetia illucens - a promising source for sustainable production of proteins, lipids and bioactive substances," *Z. Naturforsch.*, **72 (9-10)**, 352-63, 2017.

[2] Vogel H, Müller A, Heckel D G, Gutzeit H and Vilcinskas A, "Nutritional immunology: Diversification and diet-dependent expression of antimicrobial peptides in the black soldier fly Hermetia illucens," *Dev. Comp. Immunol.*, **78**, 141-8, 2017.

[3] Sun-Waterhouse D, Waterhouse G I N, You L, Zhang J, Liu Y, Ma L, Gao J and Dong Y, "Transforming insect biomass into consumer wellness foods: A review," *Food Res. Int.*, **89**, 129-51, 2016.

[4] Knidri E H, Belaabed R, Addaou A, Laajeb A and Lahsini A, "Extraction, chemical modification and characterization of chitin and chitosan," *Int. J. Biol. Macromol.*, **120**, 1181-9, 2018.

[5] Moussian B, "Recent advances in understanding mechanisms of insect cuticle differentiation," *Insect Biochem. Mol. Biol.*, **40**, 363-75, 2010.

[6] Bilandžija H, Laslo M, Porter M L and Fong D W, "Melanization in response to wounding is ancestral in arthropods and conserved in albino cave species," *Sci. Rep.*, **7**, 1-11, 2017.

[7] Vahidzadeh E, Kalra A P and Shankar A, "Melanin-based electronics: From proton conductors to photovoltaics and beyond," *Biosens. Bioelectron.*, **122**, 127-39, 2018.

[8] Brenner M and Hearing V J, "The protective role of melanin against UV damage in human skin," *J. Photochem. Photobiol.*, **84**, 539-49, 2008.

[9] Meredith P and Riesz J, "Radiative relaxation quantum yields for synthetic eumelanin," *J.*

Photochem. Photobiol., **79**, 211-6, 2004.

[10] Marei N H, El-Samie E A, Salah T, Saad G R and Elwahy A H M, "Isolation and characterization of chitosan from different local insects in Egypt," *Int. J. Biol. Macromol.,* **82**, 871-7, 2016.

[11] Waśko A, Bulak P, Polak-Berecka M, Nowak K, Polakowski C and Bieganowski A, "The first report of the physicochemical structure of chitin isolatedfrom Hermetia illucens," *Int. J. Biol. Macromol.,* **92**, 316-20, 2016.

[12] Bakulin A V, Veleshko I E, Rumyantseva E V, Levov A N, Burmistrova L A, Khismatullin R G, Kurchenko V P, Varlamov V P and Krivtsov N I, "Obtaining chitin-melanin complexes from Apis mellifera and studying the possibility of their use as as sorbents of radionuclides," *Russ. Agric. Sci.,* **5**, 48-51, 2011.

[13] Khayrova A, Lopatin S and Varlamov V, "Black soldier fly Hermetia illucens as a novel source of chitin and chitosan," *Int. J. Sci.,* **8**, 81-6, 2019.

[14] Lopatin S A, Khayrova A S, Varlamov V P and Sokolov I V, "The method of obtaining chitin from the larvae of the black soldier fly Hermetia illucens". Russian Patent 2680691 (2019).

IOP Conf. Series: Materials Science and Engineering **809** (2020) 012020 doi:10.1088/1757-899X/809/1/012020

Your Name	Title*	Research Field	Personal website
Adelya Khayrova	PhD student	Obtaining chitosan and chitosan-melanin complexes from insects and investigation of their physicochemical and biological properties	
Sergei Lopatin	PhD in Chemistry	Metal-chelate affinity chromatography, HPLC, obtaining and analysis of chitin and chitosan	
Valery Varlamov	PhD in Chemistry, Full professor	Biopolymers, chitin, chitosan, enzymes, proteins, chromatography	https://www.fbras.ru/en/about/nauchnye-podrazdeleniya/laboratoriya-inzhenerii-biopolimerov

Influence of Surface Modification of $BaTiO_3$ Nanoparticles by Sodium Oleate and Chitosan on their Optical Properties and Agglomeration in Aqueous Solutions

J A Roldan Lopez[1], L M Angelats-Silva[1], H León-León[1], R Céspedes-Vásquez[2], C W Aldama-Reyna[2], N A Emelianov[3]

[1] Antenor Orrego Private University, Trujillo 130101, Peru
[2] National University of Trujillo, Trujillo 130101, Peru
[3]Kursk State University, Kursk 305000, Russia

jroldanl@yahoo.es

Abstract. Spherical titanium barium nanoparticles with an average size about 100 nm in the tetragonal crystal phase were obtained by peroxide synthesis. To prevent their agglomeration and reduce the toxic effect, the surface of nanoparticles was chemically modified with sodium oleate and chitosan. Influence of surface modification by these compounds on agglomeration of nanoparticles in aqueous solutions, their spectral features, as well as generation of the second optical harmonic were investigated.

1. Introduction

Nowadays, significant interest in the application of ferroelectric $BaTiO_3$ nanoparticles in nanomedicine as imagine probes for different methods of microscopy, cell proliferation agents and nanotransducers for cell stimulation are observed [1, 2]. Introduction into biological systems requires stabilizing nanoparticles in biocompatible systems, usually aqueous solutions. At the same time, due to the tendency to agglomeration, as well as the leaching of metal ions, their surface is modified using different coatings [3]. Various types of coatings make it possible to obtain particles with different thickness of modifying layer, surface charge value, strength of coating adhesion. All these factors have a significant influence on the interaction of nanoparticles with cells.

Experimental results indicate that surface modification has a significant impact on the optical properties of $BaTiO_3$ nanoparticles. It is shown, that modification of the surface of $BaTiO_3$ nanoparticles by a layer of polymethylmethacrylate (PMMA) leads to an increase in the intensity of photoluminescence, which can be associated with passivation of surface defects [4]. Modification of the surface of nanoparticles of barium nanoparticles by a polystyrene layer leads to remarkable changes in phase transitions. In contrast to the volumetric samples, for these nanoparticles the orthorhombic crystal phase exists in the temperature range of 10-388 K, coexisting with the rhombohedral phase at lower temperatures and the tetragonal phase at higher temperatures. This fact has an impact on the nonlinearity of optical and dielectric properties [5]. In [6] significant influence of surfactant used in hydrothermal synthesis of $BaTiO_3$ nanoparticles was demonstrated.

Sodium oleate is widely used to stabilize inorganic nanoparticle suspensions in aqueous solutions and different organic solvents, which changes the chemical composition of the surface from hydrophilic to oleophilic [7]. Chitosan compounds are also actively used for increasing of the cytocompatibility of $BaTiO_3$ nanoparticles in cancer therapy [8]. Thus, the purpose of the work is an experimental study of the influence of the modifying layer on the optical properties, including the generation of the second harmonic in $BaTiO_3$ nanoparticles, as well as their aggregation in aqueous solutions.

2. Materials and methods

$BaTiO_3$ nanoparticles were synthesized using the peroxide method [9]. An aqueous solutions of $BaCl_2{\cdot}2H_2O$ (Merck) and $TiCl_3$ (Sigma-Aldrich) were mixed at 10 °C and then aqueous solution of H_2O_2 (J.T. Baker) was added. Then drop by drop of the NH_3 water solution was added to reach pH=9. The resulting light-yellow precipitate $BaTiO_2(O_2){\cdot}3H_2O$ was filtered, washed and dried. To obtain barium titanate, the obtained peroxide precursor was annealed for 6.5 hours at the temperature 900 °C. Modification of the surface of the obtained nanoparticles with sodium oleate was performed according to [7]. Further, 1 g of $BaTiO_3$ nanoparticles was refluxed in a 400 mL 30% aqueous solution of H_2O_2 (J.T. Baker) to enrich the surface by hydroxyl groups. Then the particles were filtered and added with solid loading of 2 wt% to a 100 mL of 0.5 wt% aqueous sodium oleate solution. Aqueous solution of SOA was prepared by mixing oleic acid (Sigma-Aldrich) and NaOH with a molar ratio of 1:1. Then mixture was heated at 90 °C for 3 h. The obtained nanoparticles were washed, filtered and dried at 90 °C during a day.

For the modification of the nanoparticle surface by chitosan, the following method was used. Chitosan (low molecular weight, Sigma-Aldrich) 0.125 g was added and diluted in 50 mL of acetic acid (CH_3COOH) and mixed up for 24 hours at 40 °C. The pH of the solution was regulated to 4.7 used 10% aqueous solution NaOH and then the solution was filtered through a membrane of 0.45 μm. Then 0.1 g of $BaTiO_3$ treated with H_2O_2 was added and mixed 4 hours at a temperature of 40 °C. As in the previous case, obtained nanoparticles were washed, filtered and dried at 90 °C during a day.

Obtained samples of nanoparticles with modified (sodium oleate or chitosan) surface were dispersed in aqua by ultrasonic treatment (60 W, 5 min.).

Structure of the synthesized nanoparticles was studied by SEM (TESCAN VEGA 3 LM) with EDX (Bruker X-Flash 6I10) and XRD (Bruker D8 ADVANCE ECO) with using Co Kα radiation line (0.17903 nm). The surface chemistry of treated nanoparticles was characterized by FTIR (Thermo Fisher Nicolet iS-50). Generation of the second harmonic by samples was researched with Nd:YAG laser 1064 nm (pulse duration 10 ns, 2.1 mJ). Dynamic light scattering analysis (Nicomp Nano Z3000) was used to evaluate the dispersibility of 1 wt.% BaTiO3 nanoparticles in aqueous solutions and for the measurement of the zeta potentials of the obtained solutions. Absorption spectra of solutions in the ultraviolet range were measured by UV/Vis spectrometer SPECORD 250 PLUS (Analytik Jena).

3. Results and discussion

SEM images of the synthesized samples demonstrate the formation of spherical nanoparticles with an average size approximately 100 nm (Figure 1).

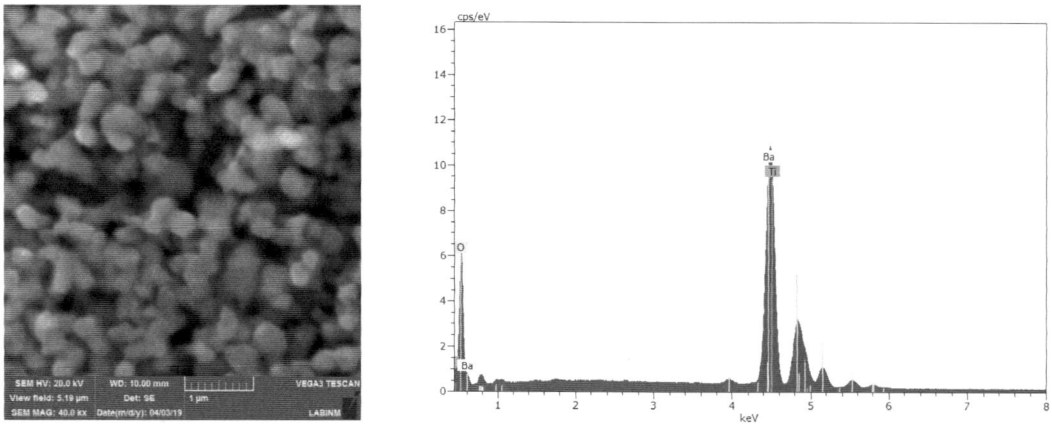

Figure 1. SEM image and EDX spectrum of $BaTiO_3$ nanoparticles obtained by peroxide synthesis.

Diffraction pattern also demonstrates the formation of nanoparticles with perovskite structure (Figure 2). Estimation of average nanoparticle size made from the widening of the peak (101) according to the Sherer equation:

$$D=0.9\lambda/\beta \cos\theta \qquad (1)$$

where D is the grain diameter, β is the half intensity width of the relevant diffraction peak, λ is the X-ray wavelength, and θ is the diffraction angle. The average particle size is estimated at 94 nm, which is close to the size estimated from the SEM images. Splitting of reflexes (001/100) and(002/200) clearly indicates the formation of the ferroelectric tetragonal crystalline phase of $BaTiO_3$.

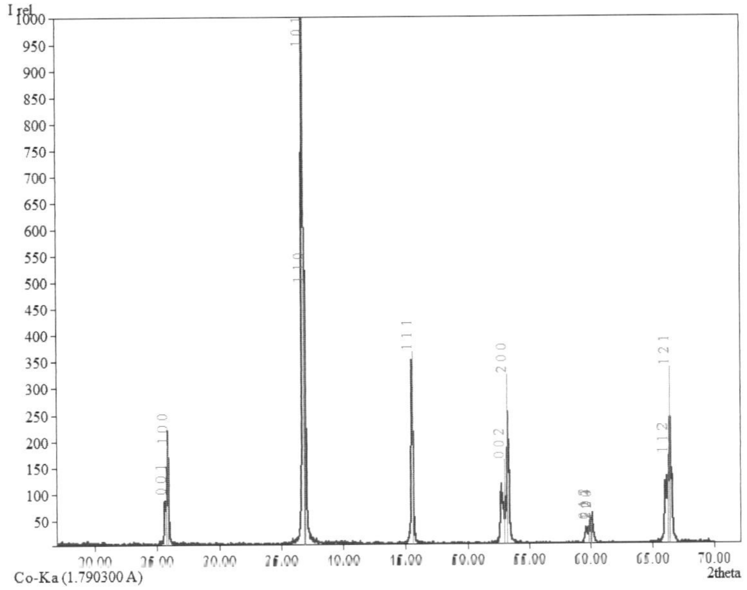

Figure 2. Diffraction pattern of $BaTiO_3$ nanopaticles obtained by peroxide synthesis.

FTIR spectrum of unmodified $BaTiO_3$ nanoparticles (Figure 3) demonstrates the presence of bands 520 cm^{-1}, caused by stretching vibrations of the Ti-O bond and 1455 cm^{-1} which is due to the stretching vibration of $-CO_3^{2-}$ associated with residual barium carbonate in $BaTiO_3$.

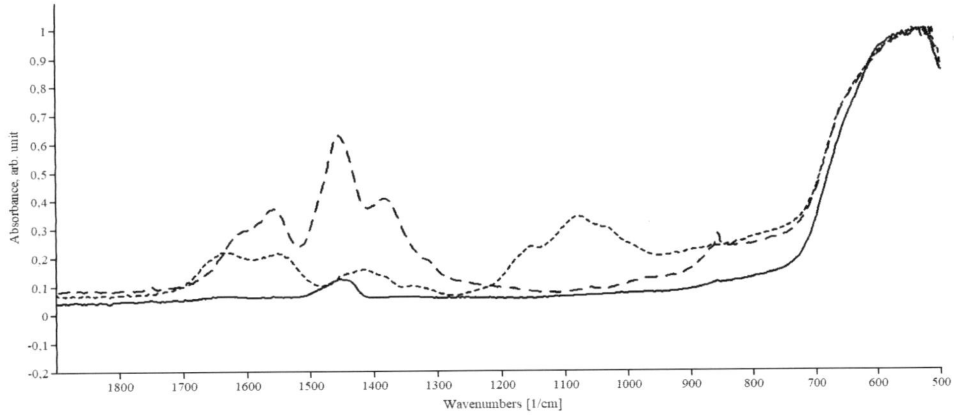

Figure 3. FTIR spectra of as-received $BaTiO_3$ nanoparticles (solid line), modified by sodium oleate (dashed line) and modified by chitosan (dotted line).

In addition, the sample of nanoparticles with modified sodium oleate surface is characterized by bands of 1384, 1557 and 1733 cm^{-1}, that indicate formation of bidentate-type bond between the surface of

nanoparticles and sodium oleate molecules [7]. The spectrum of nanoparticles modified by chitosan (Figure 3) is characterized by the presence of bands 1077, 1552 and 1632 cm^{-1}. Peak 1077 cm^{-1} corresponding to C–O–C stretching in the glucosamine. The band at 1552 cm-1corresponds to the NH_2 group and the band at 1632 cm-1 to the N–H bending vibration [10].
Both synthesized samples with unmodified surface and samples with modified surface according to the specified methods demonstrate the generation of the second harmonic with wavelength 532nm during irradiation by 1064 nm Nd:YAG laser pulses (10 ns, 2.1 mJ) (Figure 4). It should be noted that the signal intensity for samples with modified sodium oleate and chitosan surface is much lower, which can be due to the absorption of the modifying coating. In addition, modification of the surface by sodium oleate reduces the intensity of generation of the second harmonic more significantly than modification by chitosan.

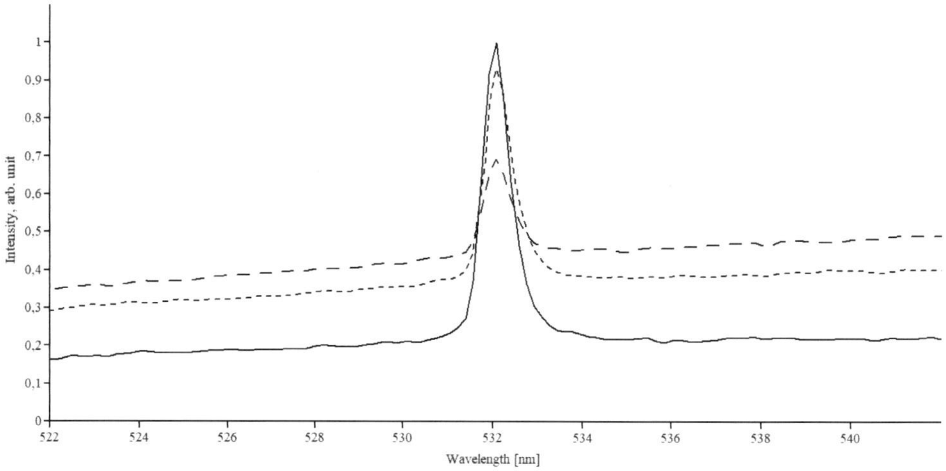

Figure 4. Second harmonic generation intensity of as-received $BaTiO_3$ nanoparticles (solid line), modified by sodium oleate (dashed line) and modified by chitosan (dotted line).

Dynamic light scattering data obtained for the dispersions of unmodified and modified nanoparticles indicates a reduction in average size for particles with modified surfaces from 216 nm for unmodified to 107 nm for sodium oleat modified and 135 nm for chitosan modified nanoparticles $BaTiO_3$ (Figure 5). It should also be noted that there is a decrease in the polydispersity of nanoparticles with modified surfaces, especially for sodium oleate modified types. It was found that for the dispersions under study, the modification of the surface with sodium oleate leads to changes in the zeta potential from 24 mV, which is close to the values obtained in [11], to -53 mV (pH=7).

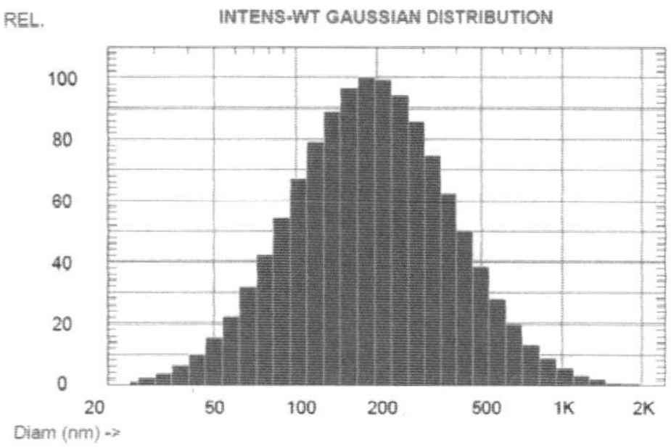

a)

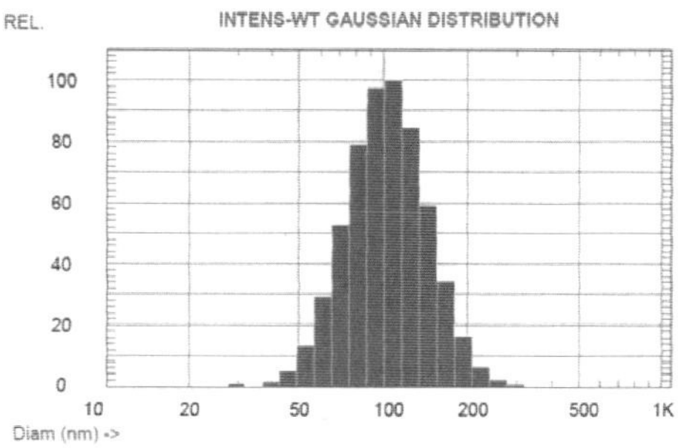

b)

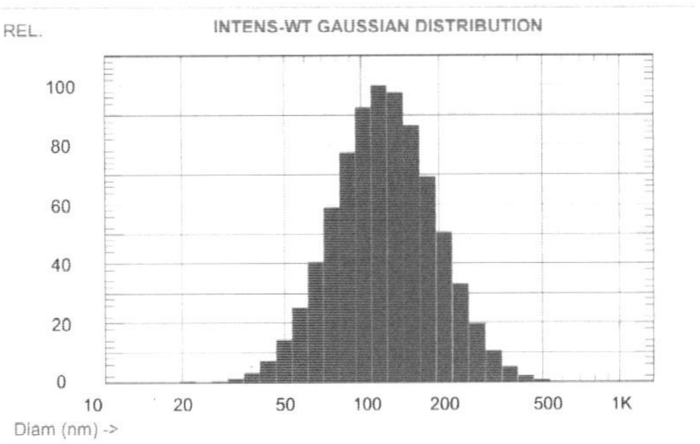

c)

Figure 5. Dynamic light scattering particle size distribution for as-received (a), modified by sodium oleate (b) and modified by chitosan (c) 1 wt.% $BaTiO_3$ nanoparticles in aqueous solutions.

At the same time, for nanoparticles modified by chitosan, the zeta potential practically does not change (26 mV).
Offset of the edge of UV absorption to shorter wavelengths for the dispersions of $BaTiO_3$ with modified surface (Figure 6) could be caused by a decrease of their polydispersity and average particle size. Thus, in these systems, the dimensional effect associated with the increase in the bandgap energy of nanostructured semiconductors is more evident [12]. Moreover, for the dispersions of modified nanoparticles another wide absorption peak around 240 nm is observed, which can be associated with different types of surface defects, including oxygen vacancies [13]. The intensity of this peak for the dispersion of sodium oleate modified nanoparticles is significantly higher than for nanoparticles with a chitosan modified surface.

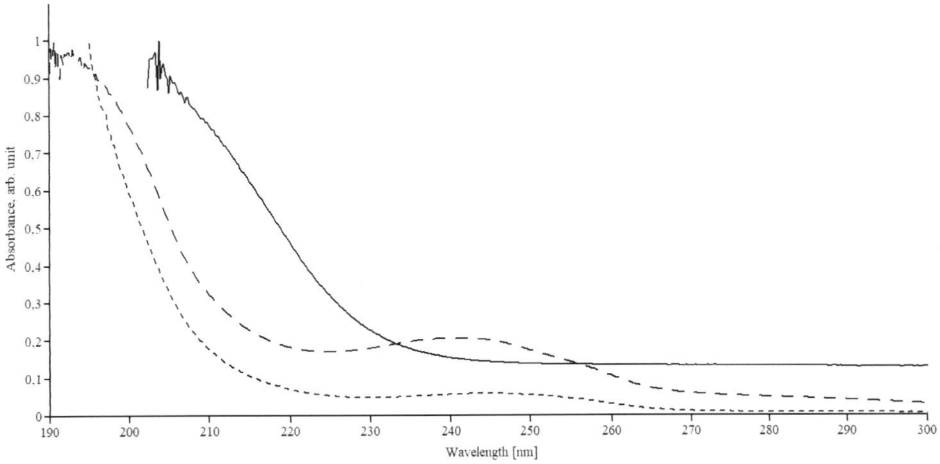

Figure 6. UV absorption spectra of 1% wt dispersions of as-received (solid line), modified by sodium oleate (dashed line) and modified by chitosan (dotted line).

4. Summary

Surface of ferroelectric $BaTiO_3$ nanoparticles with average size of 100 nm was modified by sodium oleate and chitosan. It was established that surface modification leads to a decrease in the agglomeration of nanoparticle dispersions in aqueous solutions. In this case, modification of sodium oleate, unlike chitosan, leads to a significant change in the zeta potential (from 24 to -53 mV). It is shown that the modification of the surface by these compounds also leads to a decrease in the intensity of the second harmonic generation. It is established that the modification of sodium oleate leads to a more significant reduction in the intensity of the second harmonic signal. Surface modification also leads to appearance of the broad UV absorption peak (240 nm) which can be associated with surface defects. This peak also has a significantly higher intensity for nanoparticles modified with sodium oleate than for nanoparticles modified with chitosan.

5. References

[1] Culic-Viskota J., Dempsey W.P., Fraser S.E. Pantazis P 2012 *Nature Protocols* **7** 1618-33.

[2] Marino A., Almici E., Migliorin S., Tapeinos C., Battaglini M., Cappello V., Marchetti M., de Vito G., Cicchi R., Saverio Pavone F., Ciofani G. 2019 *J. of Colloid and Interface Science* **538** 449-61.

[3] Garbovskiy Yu., Zribi O., Glushchenko A. 2012 Emerging Applications of Ferroelectric Nanoparticles in Materials Technologies, Biology and Medicine *Advances in Ferroelectrics* ed A. Pelaiz-Barranco (Rijeka: InTech) chapter 21 pp 475–497.

[4] Requena S., Lacoul S., Strzhemechny Y.M 2014 *Materials* **7** 471-83.

[5] Guo H.Z., Mudryk Ya., Ahmad M.I., Pang X.C., Zhao L., Akinc M., Pecharsky V.K., Bowler N., Linac Z.Q., Tan X 2012 *J. Mater. Chem.* **22** 23944.

[6] Vijayalakshmi R., Rajendran V 2013 *International Journal of Nanoscience* **12** 1350003.

[7] Chang S.-J., Liao W.-S., Ciou C.-J., Lee J.-T., Li C.-C 2009 *J. of Colloid and Interface Science* **329** 300–5.

[8] Ciofani G., Danti S. D'Alessandro D., Moscato S., Petrini M., Menciassi A 2010 *Nanoscale Res Lett* **5** 1093–101.

[9] Pfaff G 1988 *Z. Anorg. Allg. Chem.* **28** 76-7.

[10] Fuentes S., Dubo J., Gonzalez R., Veloso E 2015 *J. Magn. Magn. Mater.* **377** 65-9.

[11] Li J., Inukai K., Tsuruta A., Takahashi Y., Shin W. 2017 *Journal of Asian Ceramic Societies* **5** 444-51.

[12] Brus L.E 1984 *J. Chem. Phys.* **80** 4403-10.

[13] Sagdeo A., Nagwanshi A., Pokhriyal P., Sinha A.K., Rajput P., Mishra V., Sagdeo P.R 2018 *J. Appl. Phys.* **123** 161424.

Acknowledgments

The work was funded by Consejo Nacional de Ciencia, Tecnología e Innovación Tecnológica – CONCYTEC according to agreement at Sub project 072 – 2018 – FONDECYT – BM – IADT – AV.

IOP Conf. Series: Materials Science and Engineering **809** (2020) 012022 doi:10.1088/1757-899X/809/1/012022

Anatomical Kidney Model Fabrication For Ultrasound Simulation Using Direct 3D Printing

Lujie Peng[1*], Shu Wang[2]
[1]Electronics and Electrical Engineering, University of Glasgow, Glasgow, UK;
[2]Biomedical Engineering & Imaging Sciences, King's College London, London, UK

[1]Lujiepeng219@qq.com; [2]shu.1.wang@kcl.ac.uk

Abstract. With the advance of 3D printing technology showing improvements in increasing number of medical applications, there are more and more 3D kidney model printings emerging in the field. Material selection is an important aspect in 3D printing technology. Having compared many materials, Agilus30 was chosen as our final option due to its high printing accuracy, feasibility to be printed directly and desired ultrasound compatibility. ITK-SNAP 3.6.0, Seg3D and Netfabb2019 were used to segment structures in 3D medical images which were downloaded from embodi website, to smooth the model and to export the model to STL format file respectively. Eventually we printed a whole kidney in Agilus30 using Object500 Connex3 3D printer and two halves kidney in rigid resin using Anycubic 3D printer. Then we used X3-1 3Mhz probe image to get ultrasound result from the Agilus30 model in water to get outer structure and the rigid resin model in paraffin to get inner structure. After validating the result with that of a real kidney ultrasound image, we found the results are in good agreement.

1. Introduction

Traditional ultrasound compatible materials like paraffin[1], PVA-C[2], silicone[2], cannot be directly 3D printed, if they are chosen to make phantoms, a negative mold needs to be manufactured first, and then the liquid materials are poured inside. For kidney models with complicated inner structures, the mold making technique is too difficult to be implemented and the potential results will not be accurate enough. For other promising 3D printable biomaterials like hydrogel[3], Gel-wax[4], their applications are limited in small model fabrication, besides, their low printing accuracy cannot satisfy our needs. All the current 3D printable materials with high accuracy and fast printing speed, are mainly rigid materials, like PLA[5], thus, high attenuation inside these materials of the ultrasound wave and strong reflection on the material surface will result in inside structure disappearance and a lot of artefacts during imaging respectively. To have a more realistic kidney model, Agilus30 was chosen as our final option due to its high printing accuracy, feasibility to be printed directly and desired ultrasound compatibility[6]. In our paper, 6 anatomical kidney models with different pathologies were segmented and then smoothed using medical segmentation software, after model mesh generation, they are printed directly using Agilus30, and validated with ultrasound imaging about both its inner and outer structure.

2. Literature review

Rapid prototyping (RP) is a group of technologies that use computer aided design (CAD) data to produce physical size models in as short a time as possible. The construction process is carried out by "three-dimensional (3D) printing" or "addictive manufacturing". The main idea supporting this

technology is building a 3D model by creating a layer over layer. This technology began at 1980s and now have developed into a different set of technologies, which may eventually change the production process awareness of different industries [7]. 3D technology has showed its prevalence in the manufacturing industry for its utility in lowering the cost of the process to levels that make it more feasible for general manufacturing use [8]. Meanwhile, this technology is becoming increasingly important in the medical field. In the last decades, rapid prototyping has been used in a variety of medical applications including individual patient care (surgical planning, implant and tissue designing), research and as an educational and training tool. There have been many studies demonstrating the significant improvements in diagnosis and treatment due to more detailed appearance of pathological structure, increased accuracy and possibility of pre-planning [9]. For patients themselves, 3D models can help them gain a basic understanding of kidney physiology and anatomy, tumor location and characteristics and the proposed surgical procedure with its related risks, which is important to obtain patients' permission to do the renal surgery before it's arranged [10].
To plan a renal surgery, abdominal CT scans are acquired preoperatively to localize the tumor on kidney[11]. There are kidney and other surrounding organs in the image, thus accurate segmentation of the tumor and kidney is often a first step to extract features useful for the identification of disease-specific morphological differences[12]. Time-consuming as it is, manual segmentation by expert with deep knowledge of underlying anatomy and pathology is much more accurate than automatic segmentation. Even in problems where fully automatic segmentation works well, manual segmentation is still often needed to train the automatic algorithms. Thus here comes semi-automatic segmentation which combines the advantages of the two above methods to allow the human expert to make big-size objects as accurate as possible while relying on the computer to perform time-consuming repetitive tasks [13].
With the advance of 3D printing technology showing improvements in increasing number of medical applications, there are more and more 3D kidney model printings emerging in the field. At the present stage, we have with different colors and different materials [10] to show different parts and also have kidney printings with different levels of transparency and hardness in different structures to distinguish whether the area is soft or hard, normal or calcified [14]. However, the high financial cost of the models has always been the limitation to prevent the researches from producing a large amount of samples. In addition, sometimes those researches are inclined to pick those typical kidneys with seriously-deteriorated or obvious tumours to be printed in order to show the significant effect of their achievements, which makes the result not that objective.
After the kidney is printed, it will be tested to see if it is clearly visible to recognize the structure inside using ultrasound. Ultrasound imaging is designed to help surgeons see how the catheter shuttles inside the body, which is widely used in the diagnosis and image guidance of human diseases[15]. This technology is used more and more extensively because of its simple principles, relatively cheap equipment compared with tomography (CT) or magnetic resonance imaging (MRI) and the most important, non-toxic to human health (without radiation) [16].

3. Methodology

3.1 Segmentation

An original DICOM (Digital Imaging and Communications in Medicine) CT dataset with a voxel spacing of 0.6 * 0.6 * 0.6 was chosen for this project from the website Embodi3D. This dataset contains patients with a certain degree of diseases, for example, patient 1 & 2 have slightly enlarged kidney cavities while patient 3 has a huge tumor on top of the left kidney. The contrast of the images were enhanced by adjusting the software Image-J, then snake semi-automatic segmentation tool in ITK-SNAP 3.6.0 (PICSL, USA), a software in which you can set the step size for it to grow and control the evolution to see the updated 3D model in the 3D window beside (see Figure 1), was used to segment the rough kidney and the segmented model was smoothed in Seg3D using the level 2 median filter in the data filter tool column to get the uneven surface flat and give a better appearance. Then get back to the ITK-snap and export the processed model to STL format file (Stereolithography File), which is the 3D printable format file, to be ready for printing.

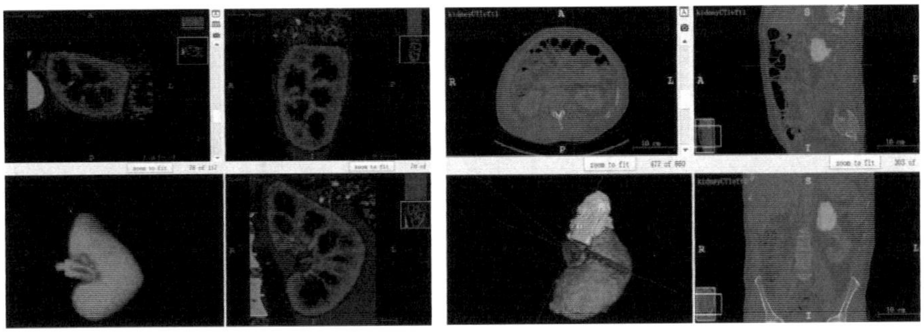

Figure 1. Segmentation for kidney in ITK-SNAP.

3.2 Agilus30 printing and clear rigid resin printing

Six models were segmented eventually. After comparing the 6 kidney models, the one can best represent was decided to print due to the high cost of printing and limitation of the budget. We chose the flexible Agilus30 because of its flexibility performance based on the reference tables.

A printing order for the kidney with an enlarged renal capsule was made to the company 'STRATASYS' due to the lack of the equipment for material Agilus30. It is printed using commercial Object500 printer (Stratasys, USA) using Aglius30 with 0.027mm layer thickness in high mix mode. Meanwhile, another half kidney was also printed for later inner structure ultrasound process using Anycubic photon printer in our lab in clear rigid resin with 0.020mm layer thickness. Both of the two printers are UV-light cured resin printers and the resin used here is cured sensitive to 405nm UV-light.

3.3 Ultrasound experiments

For the inner structure, the clear resin half kidney model was processed in ultrasound in both water and paraffin. When first using water, we were not satisfied with the result due to its dark background which made the outline of structures unclear and we thought using paraffin (see Figure 2) mimicking the tissue environment could be a better idea. Then the paraffin gel was first heated up on the magnetic stirrer and then poured around the half kidney model inside a water tank. The Philips IE33(Philips, Netherland) ultrasound machine and X3-1 3MHz probe with transmission gel were used to do the ultrasound imaging. During the imaging, the image depth was adjusted for a proper view and then the loops were acquired.

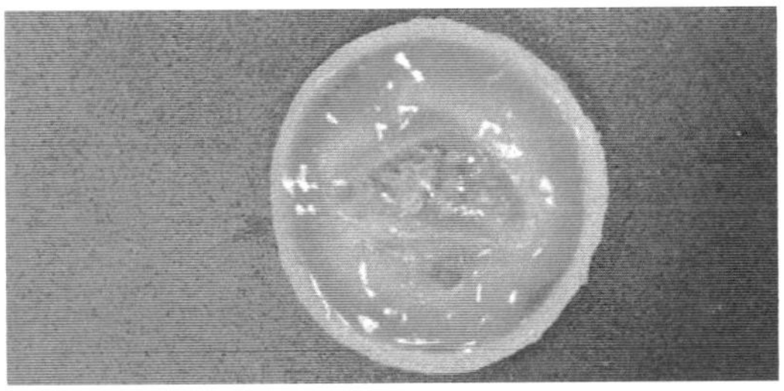

Figure 2. Ultrasound experiment for kidney model with paraffin.

The whole process which consists of segmentation, printing and ultrasound experiments is shown in the flow chart below in Figure 3.

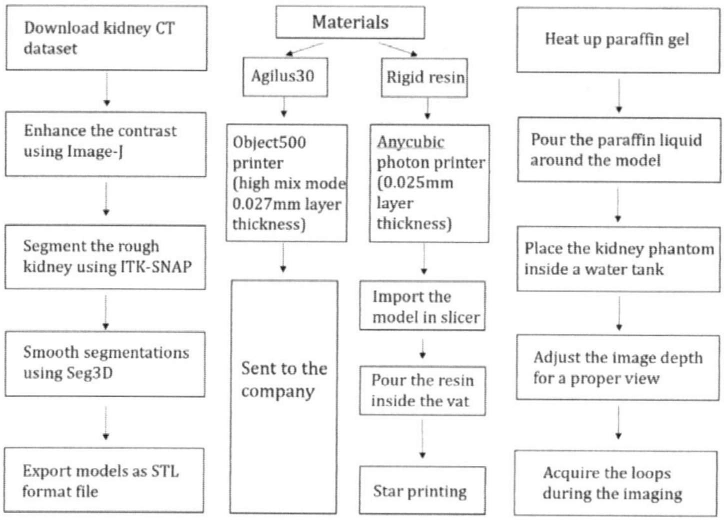

Figure 3. Flow chart for segmentation, printing and ultrasound process.

4. Results

4.1. Segmentation result

Figure 4 shows the 6 segmented kidney models from 3 different patients. Red label, blue label and green label were used to represent kidney, vessel and tumor respectively.

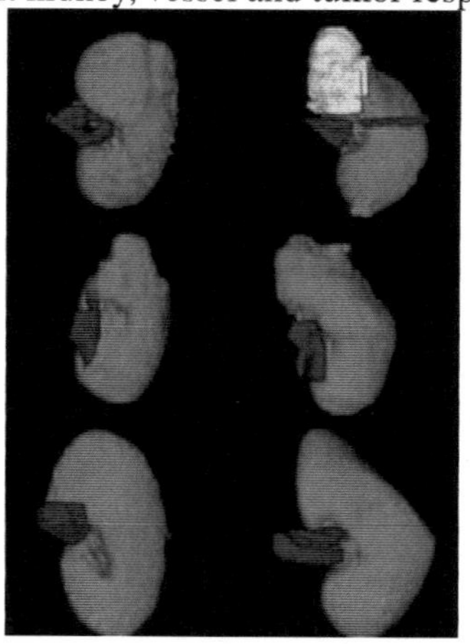

Figure 4. Six segmented kidneys from three different patients.

The first two models are from patient1 whose right kidney is in normal condition while left kidney has been invaded by a big chunk of tumour. The second two models are from patient2 whose right kidney is in normal condition while left kidney has a tumour in initial stage due to its slight deformation on the top but no visible tumour generated yet. The last two models are from patient3 whose right kidney is in normal condition while left kidney has enlarged inner chambers. This left one was also the one we thought can best present due to its clearly visible inner structures and chambers and eventually we sent this model to the company STRATASYS to be printed in Agilus30.

4.2. Printing Result

Figure 5 shows the printed real Agilus30 whole kidney model and the rigid resin separated kidney halves.

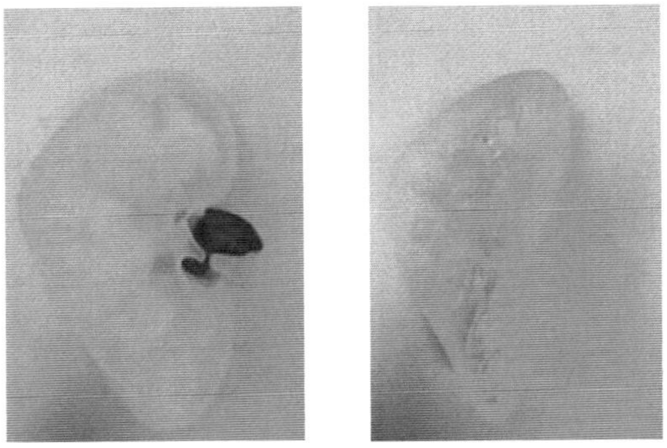

Figure 5. Agilus30 printed whole kidney & Rigid resin printed half kidney.

4.3. Ultrasound Result

The first graph of Figure 6 is the result when imaging the model with water while the latter one is with paraffin. In the one with water, the background is almost too dark to see any structure because water is lack of scattering while paraffin can be used to create an environment mimicking real tissues to give a clearer view of the inner structure.

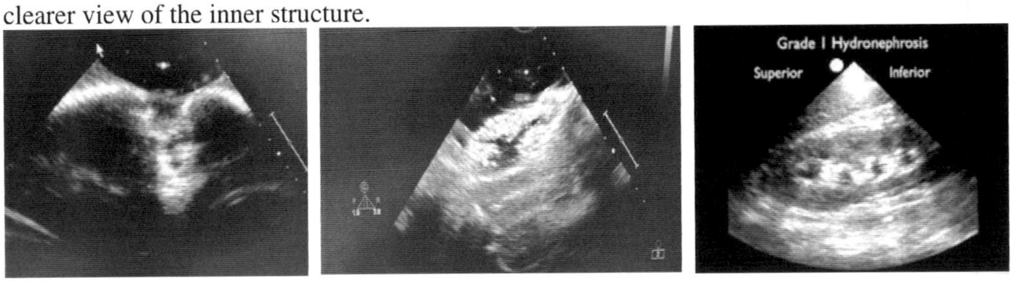

Figure 6. Inner structure with water & Inner structure with paraffin & Real kidney ultrasound image

Comparing to the real kidney ultrasound image (see the last graph of Figure 6), there is a great deal of similarity between our results of the printed kidney models and the real one and we can see there has clear outline of the inner structure.

5. Discussion and Analysis

This paper demonstrates the material Agilus30 has both high printing accuracy and desired ultrasound compatibility by printing a deformed kidney in real size in Agilus30 and did ultrasound image test on it both for outer and inner structure. The results validated that the models in our paper are in great agreement with the real human kidney.

However there still exists defect between the printed model and the real one due to the software default conditions thus deep learning based segmentation is considered to be used to get a more accurate 3D model.

6. References

[1] Efthymios Maneas, Wenfeng Xia, Daniil I Nikitichev, Batol Daher, Maniragav Manimaran, Rui Yen J Wong, Chia-Wei Chang, Benyamin Rahmani, Claudio Capelli, Silvia Schievano, Gaetano Burriesci, Sebastien Ourselin, Anna L David, Malcolm C Finlay, Simeon J West, Tom Vercauteren, and Adrien E Desjardins, Anatomically realistic ultrasound phantoms using gel wax with 3D printed moulds, *Phys Med Biol.* 2018 Jan; **63**(1): 015033.

[2] Deirdre M King, Carmel M Moran, John D. Mcnamara, Jacinta E Browne, Development of a Vessel-Mimicking Material for use in Anatomically Realistic Doppler Flow Phantoms, May 2011,*Ultrasound in medicine & biology* **37**(5):813-26, DOI: 10.1016/j.ultrasmedbio.2011.02.012

[3] Vivian K. Lee and Guohao Dai, Printing of Three-Dimensional Tissue Analogs for Regenerative Medicine, *Ann Biomed Eng*. 2017 Jan; **45**(1): 115–131

[4] Efthymios Maneas, Wenfeng Xia, Olumide Ogunlade, Adrien E. Desjardins, Gel wax-based tissue-mimicking phantoms for multispectral photoacoustic imaging, January 2018*Biomedical Optics Express* **9**(3) DOI: 10.1364/BOE.9.001151

[5] James Shin, Ranjit S. Sandhu, George Shih, Imaging Properties of 3D Printed Materials: Multi-Energy CT of Filament Polymers, *J Digit Imaging*. 2017 Oct; **30**(5): 572–575.

[6] APPLICATION CUSTOMER STORY: The Jacobs Institute, Agilus30™ Improves Performance of Vascular Models

[7] Ms. Sahana V.W., Dr. G.T. Thampi, 3D Printing Technology in Industry, *Proceedings of the Second International Conference on Inventive Systems and Control* (ICISC 2018)

[8] Hakmin Lee , Ngoc Ha Nguyen , Sung Il Hwang , Hak Jong Lee , Sung Kyu Hong , Seok-Soo Byun , Personalized 3D kidney model produced by rapid prototyping method and its usefulness in clinical applications, *Int Braz J Urol*. 2018; **44**: 952-7

[9] Rengier F, Mehndiratta A, von Tengg-Kobligk H, Zechmann CM, Unterhinninghofen R, Kauczor HU, et al. 3D printing based on imaging data: review of medical applications. *Int J Comput Assist Radiol Surg*. 2010;5:335-**41**.

[10] Jean-Christophe Bernhard, Shuji Isotani, Toru Matsugasumi, Vinay Duddalwar, Andrew J. Hung, Evren Suer, Eduard Baco, Raj Satkunasivam, Hooman Djaladat, Charles, Metcalfe, Brian Hu, Kelvin Wong, Daniel Park, Mike Nguyen, Darryl Hwang, Soroush T. Bazargani, Andre Luis de Castro Abreu, Monish Aron, Osamu Ukimura, Inderbir S. Gill, Personalized 3D printed model of kidney and tumor anatomy: a useful tool for patient education, *World J Urol* DO**10**.1007/s00345-015-1632-2

[11] Ivan Figueroa-Garcia , Jean-Marc Peyrat , Ghassan Hamarneh and Rafeef Abugharbieh, Biomechanical kidney model for predicting tumor displacement in the presence of external pressure load, 31 July 2014, 978-1-4673-1961-4

[12] Annemie Ribbens, Jeroen Hermans, Frederik Maes, Senior Member, IEEE, Dirk Vandermeulen, Paul Suetens, Member, IEEE, and Alzheimer's Disease Neuroimaging Initiative, Unsupervised Segmentation, Clustering, and Groupwise Registration of Heterogeneous Populations of Brain MR Images, *IEEE TRANSACTIONS ON MEDICAL IMAGING,* VOL. **33**, NO. 2, FEBRUARY 2014

[13] Paul A. Yushkevich , Yang Gao , Guido Gerig, ITK-SNAP: an interactive tool for semi-automatic segmentation of multi-modality biomedical images, **10**.1109/*EMBC*.2016.7591443, 18 October 2016

[14] Gaétan Denizet, Paul Calame, Thomas Lihoreau, François Kleinclauss, Sébastien Aubry, 3D multi-tissue printing for kidney transplantation, **10**.21037/qims.2018.10.16

[15] Aaron Fenster ; Jeff Bax ; Hamid Neshat ; Derek Cool ; Nirmal Kakani ; Cesare Romagnoli, 3D ultrasound imaging in image-guided intervention, 26-30 Aug. 2014, 978-1-4244-7929-0

[16] Xishi Huang, John Moore, Gerard Guiraudon, Douglas L. Jones, Daniel Bainbridge, Jing Ren, and Terry M. Peters, Dynamic 2D Ultrasound and 3D CT Image Registration of the Beating Heart, *IEEE TRANSACTIONS ON MEDICAL IMAGING*, VOL. **28**, NO. 8, AUGUST 2009

IOP Conf. Series: Materials Science and Engineering **809** (2020) 012023 doi:10.1088/1757-899X/809/1/012023

Effect of pH on Swelling Profile of Glutaraldehyde Crosslinked *Adenanthera pavonina* L. Galactomannan

L A Cavalcante[1, a], B S Batista[2, b], A P G Fonseca[3, c], A J D Balteiro[3, d], F M B Amaral[4,8, e], F Mendes[5,6,7, f], A A M Macêdo[1, g].

[1]Federal Institute of Education, Science and Technology of Maranhão, IFMA, Research Laboratory, Imperatriz, Brazil; [2]Federal University of Maranhão, UFMA Advanced Campus, Imperatriz, Brazil;[3]Polytechnic Institute of Coimbra, ESTESC Coimbra Health School, Department Biomedical Laboratory Sciences, Coimbra, Portugal; [4]Polytechnic Institute of Coimbra, ESTESC Coimbra Health School, Department Complementary Sciences, Coimbra, Portugal;[5]Polytechnic of Coimbra, ESTeSC, DCBL, Coimbra, Portugal; [6]CNC. IBILI Consortium/ Center for Innovative Biomedicine and Biotechnology (CIBB), University of Coimbra, Coimbra, Portugal; [7]Biophysics Institute, Coimbra Institute for Clinical and Biomedical Research (iCBR) area of Environment Genetics and Oncobiology (CIMAGO), Faculty of Medicine, University of Coimbra, Coimbra, Portugal.[8]I3N - University of Aveiro, Aveiro, Portugal.

[a]l.a.cavalcante@outlook.com, [b]batistas.bea@gmail.com,
[c]paula_fonseca@estescoimbra.pt, [d]balteiro@estescoimbra.pt,
[e]filipe.amaral@estescoimbra.pt, [f]fjmendes@estescoimbra.pt and
[g]anaangellica@yahoo.com.br.

Abstract. Galactomannans are neutral natural polymers isolated from legume endosperm and with wide use in the biomedical field. As a consequence, there is a need to investigate materials for their properties in order to better understand and apply them. In this study, we analyzed the influence of pH medium on the crosslinking process of galactomannan and glutaraldehyde, changing the pH medium from 3 to 7, using a strong base. The Fourier Transform Infrared Spectrum (FTIR) and the swelling profiles of the glutaraldehyde crosslinked galactomannan were evaluated. We have found that the glutaraldehyde-galactomannan crosslinking process is highly sensitive to the variation of pH medium. According to our results, best swelling capacity was achieved for samples prepared at pH=5, suggesting that this pH medium conditions is suitable for possible production of hydrogels prepared from glutaraldehyde-galactomannan.

1 Introduction

Polymers are macromolecules with physical properties that depend on factors such as chemical composition, mechanical configuration and length of the polymer chain [1]. Polysaccharides are polymers joined by glycosidic bonds in the α or β forms. They are rich in OH groups, able to form intra or intermolecular hydrogen bridges, which favors the control of solubility [2,3].

Galactomannans are polysaccharides composed of (Figure 1) D-mannose and D-galactose [4]. They present a linear skeleton consisting of D-mannopyranose β - (1 → 4). According to Dea and Morrison

[5] there is substitution of D – mannopyranose (C - 6) hydroxyl for D – galactopyranose α - (1 → 6) in varying degrees of substitution. What differs in galactomannans is the mannose / galactose ratio as well as the species-dependent distribution throughout the skeleton of mannopyranose.

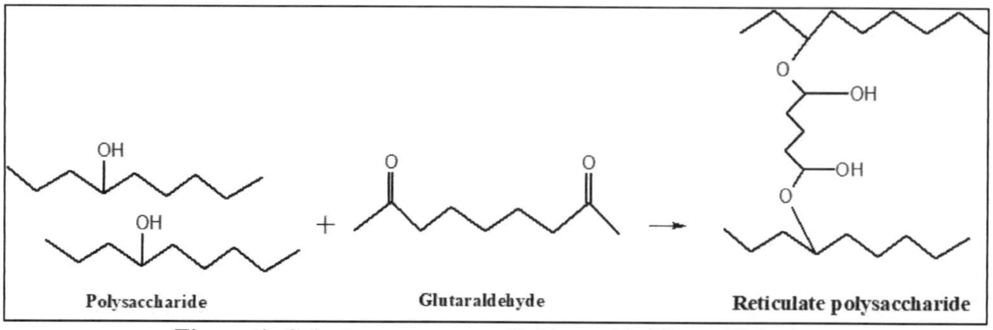

Figure 1. Galactomannan crosslinking with Glutaraldehyde.

In polymers, the crosslinking process increases mechanical and chemical resistance as well as decreases deformation. In the production of hydrogels, the crosslinking agents must have molecules with low molar mass and reactive functional groups to allow intercrossing between polymer chains [6]. In this study, *Adenanthera pavonina* L. galactomannan was crosslinked with glutaraldehyde (0.3 mol/L) using different pH medium, ranging from 3 to 7, in order to study its influence on the crosslinking process. Fourier transform infrared spectroscopy was used to identify existing functional groups and changes between different samples according to the pH variation of the preparation medium. Also the determination of the swelling degree was performed for all the samples, considering the importance of this characteristic for future applications of this biomaterials as hydrogels.

2 Methodology

2.1 Materials

Adenanthera pavonina L. seeds were collected in Imperatriz, Maranhão, in January 2019. They were harvested, selected, washed and kept in a cool place until use. In addition, for the crosslinking of galactomannan (Gal), glutaraldehyde (GA) chemical reagent was used.

2.2 Methods

2.2.1 Galactomannan extraction from Adenanthera pavonina L. seeds. In order to obtain the Galactomannan, the process started with the collection of the seeds of *Adenanthera pavonina* L., followed by the selection and cleaning of the seeds. Then they will pass through a heating and swelling stage. Finally, the seeds were manually separated from the endosperm and lyophilized.

2.2.2 Galactomannan crosslinking. A solution of 0.3 mol/L of Glutaraldehyde was prepared and its pH was adjusted with acetic acid. The solution was added to the fine powder of galactomannan and stirred for 24 hours. After stirring, the solution was placed on a petri dish for formatting. Three different galactomannan films were obtained and identified as G2G03P3, G2G03P5 and G2G03P7, according to the respective pH medium (3, 5 and7).

2.2.3 Determination of the swelling degree of galactomannan. The samples' swelling was obtained dipping the samples G2G03P3; G2G03P5 and G2G03P7 in 50 mL of distilled water. The immersion time periods were 20, 40, 60 and 80 minutes. After this process excess water was removed and the sample masses were weighted. Equation 1 was used to determine the swelling degree.

$$GI = [(M - m)/m] \times 100 \tag{1}$$

Where: GI = swelling degree; M = swollen mass and m = initial mass.

IOP Conf. Series: Materials Science and Engineering **809** (2020) 012023 doi:10.1088/1757-899X/809/1/012023

2.2.4 Infrared characterization with fourier transform. FTIR spectrum was obtained from the Bruker Fourier transform spectrometer, model Vertex 70V. Potassium Bromide (KBr) pellet technique was used and each measure was performed taking 64 scans.

3 Results and discussion

3.1 Infrared Fourier Transform as a function of pH

The FTIR spectrum allows the identification of chemical groups. The transmittance data of galactomannan in the infrared region is shown in Figure 2. The band between 810-820 cm^{-1} refers to the α-D-galactose linking unit, the band between 870-880 cm^{-1} the β-D-manose linkage unit [7,8], while the band between 950 cm^{-1} - 970 cm^{-1} was attributed to the axial deformation (C − OH) of C-4 [9]. At 1027cm^{-1}, the band may be related to the vibrational torsion of the CH_2. Already the band at 1160 cm^{-1} can be attributed to the module connection vibrational angular voltage δ(C—O) due to pyranose ring [9,10]. In addition, the region between 1350 cm^{-1} and 1450 cm^{-1}, attributed to symmetrical deformations of groups CH_2 e COH [7,8]. While the wavelength at 1640 cm-1 presents bands on account of the vibration of (O-H) of adsorb and stretch water molecules (C-O) of the carboxylate group (-COO-)[11]. The band centered at 1740 cm^{-1} is associated with carbonyl C=O group. Also between 2930 cm^{-1} and 2940 cm^{-1} band can be referred to stretch C-H of the CH_2 grouping and the 3420 cm^{-1} to the vibrational stretching of the group O-H [7].

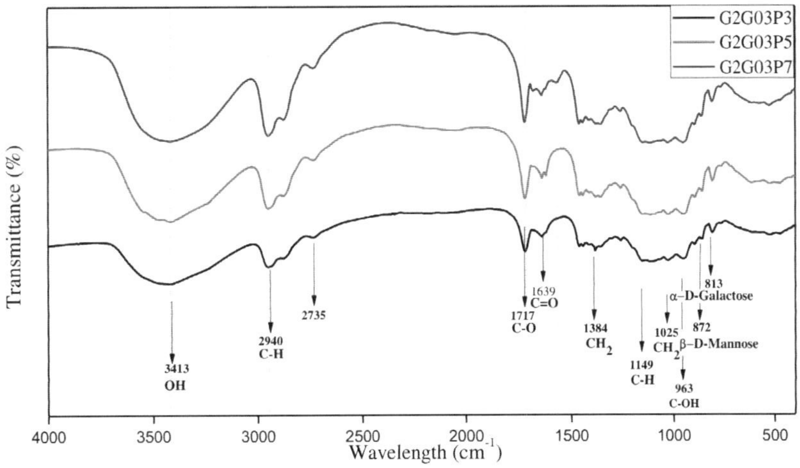

Figure 2. FTIR spectra of reticulated Galactomannan in glutaraldehyde 0.3 mol/L, in pH 3, pH 5 e pH 7.

3.2 Swelling degree as a function of pH

Figure 3 shows the swelling profile of glutaraldehyde crosslinked *Adenanthera pavonina* L. galactomannan at pH 3, 5 and 7; It is observed that the sample G2G03P5 presented a higher swelling degree compared to G2G03P3, for a dipping period of 20 min. It is also noted that lower pH (higher H^+ concentration) medium correspond to a decrease of the swelling degree, which should be caused by protonation of carboxylic groups in the polysaccharide structure, leading to more interactions between carboxylic and hydroxyl groups, forming more hydrogen bonds. This effect explains the decrease of free volume accessible to water molecules, lowering water penetration [12–15]. Meanwhile, the sample G2G03P7 has no swelling capacity at all since it solubilized.

In addition, crosslinking hinders the mobility of the polymer chain, decreasing solubility of the polysaccharide[16]. The reticulating agent (glutaraldehyde) favors an increase of the crosslinking degree, affecting hydrogels structure. This phenomena, directly related with an higher degree of ramifications of the polymer and folding of polymer chains allied to eventual reticulation points [17], greatly affects water absorption capacity.

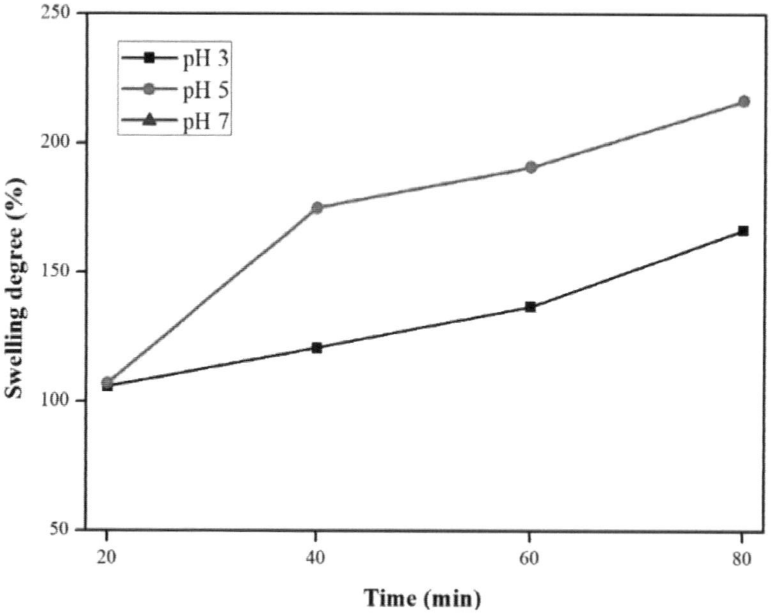

Figure 3. Swelling degree (%) for increasing periods, from 20 min to 80 min using 0.3 mol / L glutaraldehyde crosslinked galactomannan at pH=3 and pH= 5.

4 Conclusion

This research analyzed the influence of pH medium on the crosslinking process of galactomannan and glutaraldehyde. Three different pH preparation medium were used (pH=3, 5 and 7). The FTIR response of the hydrogels that resulted from crosslinking process revealed that the bands of the OH and C=O functional groups presented an increasing intensity associated to the higher pH medium. Meanwhile, our results point out that to a higher pH medium, during crosslinking, corresponds a higher swelling degree, suggesting a high sensitivity of this parameter to pH variation. Finally, considering its simplicity, this preparation method, can be easily scalable and be applied in the production of hydrogels for commercial purposes.

5. Reference

[1] Bajpai P. Polymer Chemistry. Biermann's Handb Pulp Pap 2018:373–80. doi:10.1016/B978-0-12814238-7.00018-0.
[2] Rinaudo M. Main properties and current applications of some polysaccharides as biomaterials. Polym Int 2008;57:397–430. doi:10.1002/pi.2378.
[3] Cunha PLR, Paula RCM d., Feitosa JPA. Purification of guar gum for biological applications. Int J Biol Macromol 2007;41:324–31. doi:10.1016/j.ijbiomac.2007.04.003.
[4] Kontogiorgos V. Galactomannans (Guar, Locust Bean, Fenugreek, Tara). Elsevier; 2018. doi:10.1016/B978-0-08-100596-5.21589-8.
[5] Dea ICM, Morrison A. Chemistry and interactions of seed galactomannans. Adv Carbohydr Chem Biochem 1975;31:241–312. doi:10.1016/S0065-2318(08)60298-X.
[6] De Azeredo HMC, Rosa MF, De Sá M, Souza Filho M, Waldron KW. The use of biomass for packaging films and coatings. Adv. Biorefineries, Elsevier; 2014, p. 819–74. doi:10.1533/9780857097385.2.819.

[7] Figueiró SD, Góes JC, Moreira RA, Sombra ASB. On the physico-chemical and dielectric properties of glutaraldehyde crosslinked galactomannan-collagen films. Carbohydr Polym 2004;56:313–20. doi:10.1016/j.carbpol.2004.01.011.
[8] Yuen SN, Choi SM, Phillips DL, Ma CY. Raman and FTIR spectroscopic study of carboxymethylated non-starch polysaccharides. Food Chem 2009;114:1091–8. doi:10.1016/j.foodchem.2008.10.053.
[9] Wang J, Somasundaran P. Study of galactomannose interaction with solids using AFM, IR and allied techniques. J Colloid Interface Sci 2007;309:373–83. doi:10.1016/j.jcis.2006.10.086.
[10] Kapoor VP. A galactomannan from the seeds of Delonix regia. Phytochemistry 1972;11:1129–32. doi:10.1016/S0031-9422(00)88465-4.
[11] Rodrigues SC, Salgado CL, Sahu A, Garcia MP, Fernandes MH, Monteiro FJ. Preparation and characterization of collagen-nanohydroxyapatite biocomposite scaffolds by cryogelation method for bone tissue engineering applications 2012:1–15. doi:10.1002/jbm.a.34394.
[12] Khan S, Ranjha NM. Effect of degree of cross-linking on swelling and on drug release of low viscous chitosan/poly(vinyl alcohol) hydrogels. Polym Bull 2014;71:2133–58. doi:10.1007/s00289-014-1178-2.
[13] Clara I, Lavanya R, Natchimuthu N. pH and temperature responsive hydrogels of poly(2-acrylamido-2-methyl-1-propanesulfonic acid-co-methacrylic acid): Synthesis and swelling characteristics. J Macromol Sci Part A Pure Appl Chem 2016;53:492–9. doi:10.1080/10601325.2016.1189282.
[14] Pourjavadi A, Mahdavinia GR. Superabsorbency, pH-sensitivity and swelling kinetics of partially hydrolyzed chitosan-g-poly(acrylamide) hydrogels. Turkish J Chem 2006;30:595–608.
[15] Al-anbakey AM. EFFECT OF pH ON SWELLING PROPERTIES OF COMMERCIAL POLYACRYLIC 2014.
[16] Ugwu OB, Barminas JT, Nkafamiya II, Akinterinwa A. Synthesis and Characterization of Crosslinked Hydrogel of Konkoli (<i>Maesopsis eminii</i>) Grafted Polymethylacrylamide for a Preliminary Study as an Insulin Delivery System. J Biosci Med 2016;04:36–47. doi:10.4236/jbm.2016.411005.
[17] Adamawa Y. Swelling behaviour of konkoli (maesopsis eminii) galactomannan hydrogels . 2011;2:78–86.

Acknowledgements
The authors are grateful to the Foundation for Research and Scientific and Technological Development of Maranhão - FAPEMA for financial support through EDICT FAPEMA No. 36/2017 International Cooperation Process COOPI-07965/17 and the Scientific and Technological Development Council (CNPq), a Foundation for Science and Technology (FCT), Portugal (Strategic Projects UID/NEU/04539/2013 and UID/NEU/04539/2019) and COMPETE-FEDER (POCI-01-0145-FEDER-007440) Funded by National Funds via FCT (Foundation for Science and Technology) through the Strategic Project UIDB/04539/2020 and UIDP/04539/2020 (CIBB).

IOP Conf. Series: Materials Science and Engineering **809** (2020) 012024 doi:10.1088/1757-899X/809/1/012024

Optical Simulation of Triangular Wired Grid Polarizer for Far Ultraviolet Applications

Zhuan Zhao[1,a], Enrico Tessarolo[1,2,b], Alain J.Corso[1,3,c] and Maria G.Pelizzo[1,3,d]

[1]University of Padova, Department of Information Engineering, Via Gradenigo 6/B, 35131, Padova, Italy
[2]Centre of Studies and Activities for Space, University of Padova, Via Venezia 15, 35131 Padova, Italy
[3]National Research Council of Italy, Institute for Photonics and Nanotechnologies e Padova, Via Trasea 7, 35131, Padova, Italy

[a]zhaozhuan2008@gmail.com, [b]enrico.tessarolo1984@gmail.com, [c]alain.corso@pd.ifn.cnr.it, [d]pelizzo@dei.unipd.it

Abstract. Polarimetric observations in the far ultraviolet (FUV) are very useful for us to understand the role played by the magnetic field of the coronal plasma in the energy transfer process from the inner parts of the sun to the outer space. To observe these processes there are various key spectral lines in the FUV, from which H I Lyman α (121.6 nm) is the strongest one. We designed and optimized a triangular wire grid polarizer (WGP) based on Al/MgF2 multilayer coatings to obtain high polarization with a simultaneous high Rs. As a result, at incidence angle of 50° and 60° whereas Rp was minimized, this model have a 99.98% and 99.99% degree of polarization with a Rs of 0.32 and 0.69 respectively.

1. Introduction

Polarizers for FUV applications have been studying for many years [1-4]. FUV includes many important spectral lines for astrophysics, solar physics, and atmospheric physics, one of the key FUV spectral line is H I Lyman at 121.6 nm.

Al/MgF2 multilayer thin films are commonly used for polarizers at this wavelength because of the high reflectance of the Al, lower absorption of MgF2. Therefore, many scholars have conducted research on it in FUV applications. Kim et al. [5] achieved 87.95% reflectance for Rs and 0.43% for Rp with a 99.03% degree of polarization by means of a MgF2/Al/MgF2 three layer structure on an opaque thick film of Al as the substrate. Larruquert et al. [6] designed a polarizer based on Al/ MgF2 multilayer coatings resulted in efficient polarizers at 121.6nm, with Rp~0.01-0.017 and Rs~0.69-0.725. Bridou et al. [7] based on Al/ MgF2 multilayer coatings, obtained 69% reflectance for Rs at the incident angle of maximum polarization. However, all of their maximum polarization angles are closed to glancing angle. Therefore, an efficient nano-optical wire grid polarizer (WGP) could be useful to achieve high polarization below glancing angle because of large achievable element sizes (wafer size), compactness (wafer thickness), and large acceptance angles [8].

In this paper, we designed a triangular WGP based on a Al/MgF2 multilayer thin films to shape light in its propagation behavior in FUV application. As is shown in Figure 1, where h, w_1 and w is height, width and period of the WGP. Such application needs to be evaluated by using the polarization degree parameter defined as

$$P=\frac{(Rs-Rp)}{(Rs+Rp)} \quad (1)$$

where Rs and Rp are the reflectances in S and P polarization respectively. This parameter has to be more than 99.6% and a Rs more than 0.2. Through the simulation and analysis, revealed that this kind of polarizers are promising.

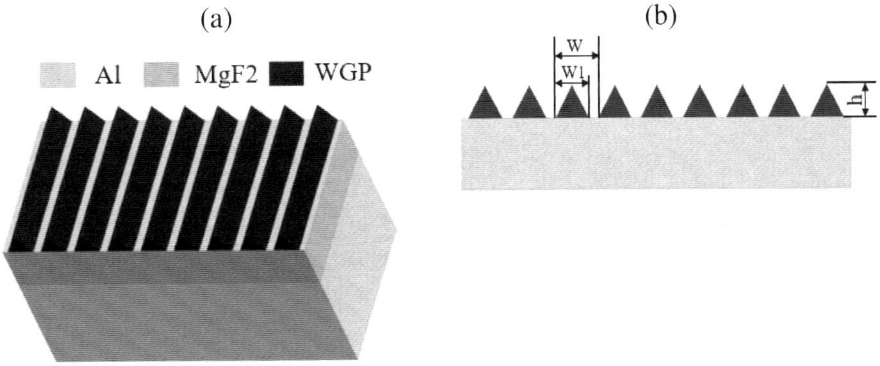

Figure 1. (a) 3D theoretical model, (b)2D theoretical model.

2. Material and structure requirements

In our model, 150 nm for the thickness of Al has been chosen, because this value is enough to reflect all the incoming radiation at the target wavelength 121.6 nm. In order to choose the optimal thickness for MgF_2, we set the thickness of MgF_2 as a variable parameter in the IMD simulations [9]. We selected the working angle of 45°, which is a value useful for folding mirrors in optical systems. The results are reported in Figure 2, where the reflectance at the working angle has been plotted versus MgF_2 thickness. We selected the MgF_2 thickness of 65 nm because it produces a constructive interference and it guarantee a good protection for the Al oxidation.

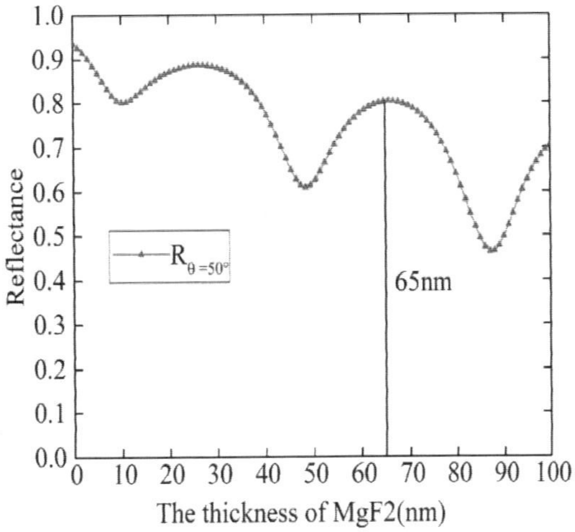

Figure 2. The reflectance versus MgF_2 thickness.

IOP Conf. Series: Materials Science and Engineering **809** (2020) 012024 doi:10.1088/1757-899X/809/1/012024

Moreover, on top of the Al\$MgF_2$ bi-layer a TiO_2 layer was selected as absorber. This topmost layer is the starting point to be etched for the realization of absorbing nano-structures on-top of the high-reflective Al\$MgF_2$ bi-layer. The thickness of this layer as well as all the parameters required to describe the triangular nano-structure (i.e. width, height and period) were carried out by using the wave optics module of COMSOL Multiphysics software [10]. Due to the symmetry of our structure, we use a 2-D modeling of the structure. The optical constants of TiO_2 was taken from literature [8] whereas for the all other materials were taken from IMD database (see Table 1) .

Table 1. Optical constants.

Material	n	k
TiO_2	1.1222	0.9157
Al	0.0423	1.1404
MgF_2	1.7047	0.0180
Air	1	0

The geometry of the model used in the simulations are presented in Figure 3. The top edge of the air domain was set up with incident port and the bottom edge of the Al domain was set up with exit port, the periodic Floquet's boundary condition established on the corresponding left and right boundaries of the model.

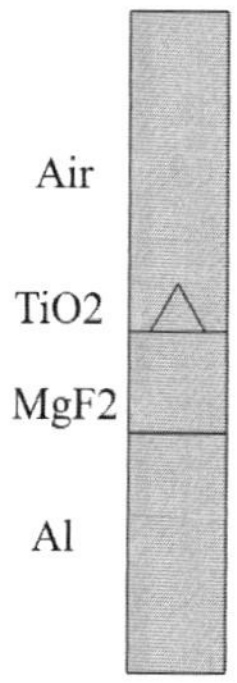

Figure 3. Geometry model.

Futher more, the incident angle is a variable, so four parameters sweep at the same time.

3. Results

After the optimization process, two different structures for WGP ensuring good performance have been found (see Table 2).

Table 2. Optimal parameters

	optimal parameters	Rs	P
Structure A	h=51nm, w_1=36nm, w=68nm, θ=50°	0.32	99.98%
Structure B	h=29nm, w_1=35nm, w=64nm, θ=60°	0.69	99.99%

The S and P reflectance versus incidence angle of each structure is reported in Figure 4. The best result is obtained for Structure B which shows the highest polarization selection (P=0.9999) as well as a highest reflectance at 121.6 nm.

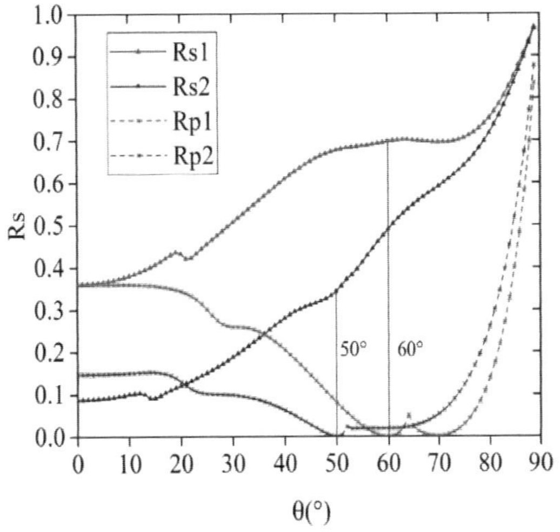

Figure 4. Reflectance versus different incident angles.

For both optimized structures, the analysis of the sensitivity between the three free parameters and the final optical performance has been carried out. This sensitivity was evaluated by swiping one parameter at a time and leaving the others at the nominal values of Table 2. The dependence graphs are reported in Figure 5. From the analysis emerged that the structure reflectance is little affected by the optimization parameters whereas the polarization degree is highly conditioned by these parameters. For instance, a polarization degree greater than 99.6% is preserved for deviation of just several nanometers for both the height (h) and the triangle base width (w1). On the other hand, the structure periodicity results to be very decisive on the polarization, requiring a tolerance that seems to be lower than 1 nm.

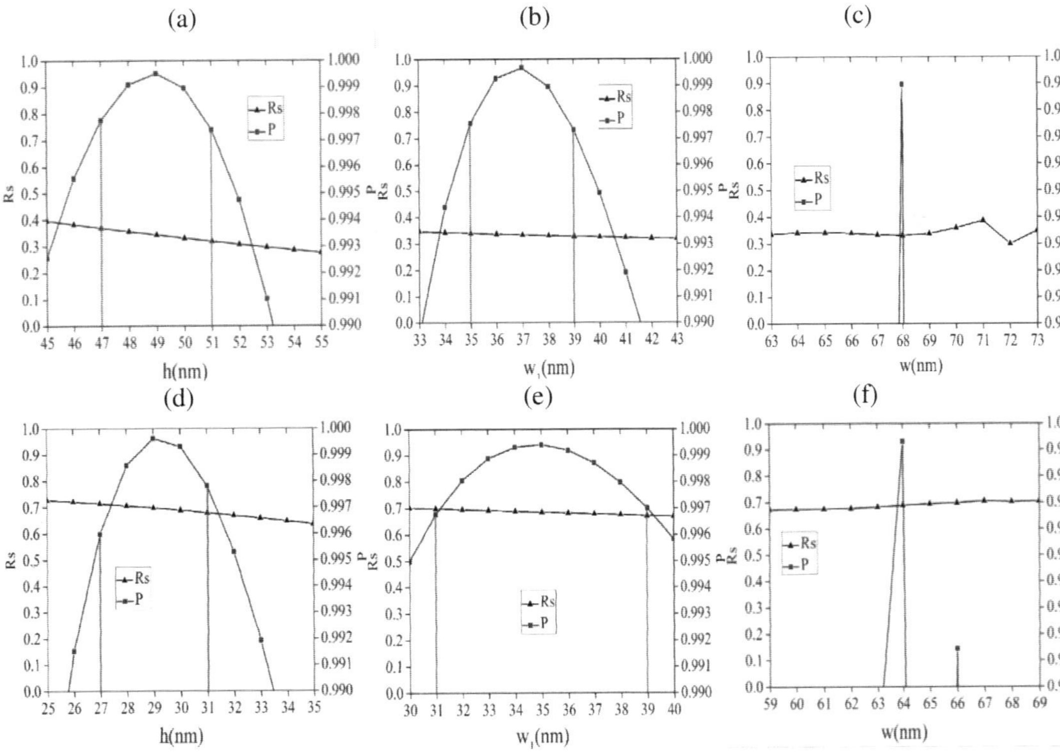

Figure 5. (a), (b) and (c) are P and Rs curve versus height, width and period of Structure A. (e), (f) and (g) is P and Rs curve versus heigh, width and period of Structure B.

IOP Conf. Series: Materials Science and Engineering **809** (2020) 012024 doi:10.1088/1757-899X/809/1/012024

4. Summary
We designed and simulated a WGP based on triangular nanostructure for FUV application. The reported results suggest a good performance in both Rs and P, especially when incident angle equal to 60°, the reflectance can be reached 0.69, and we also discuss the tolerance of key parameters. These can provide much information for our future manufacturing.

5. References
[1] J. Kim, M. Zukic, M.W. McColgan, J.H. and D.G. Torr: X-Ray and Ultraviolet Spectroscopy and Polarimetry. Vol. 2283. International Society for Optics and Photonics(1994).
[2] J. Kim, M. Zukic, M.W. McColgan, and D.G. Torr: Multilayer and Grazing Incidence X-Ray/EUV Optics for Astronomy and Projection Lithography. Vol. 1742. International Society for Optics and Photonics(1993)
[3] J.I. Larruquert, A.M. Malvezi, A. Giglia, J.A. Aznarez, L. Rodriguez-de Marcos and S. Fineschi: EUV and X-ray Optics: Synergy between Laboratory and Space IV. Vol. 9510. International Society for Optics and Photonics(2015)
[4] J.I. Larruquert and R.A. M Keski-Kuha: Applied optics 38.7, 1231-1236 (1999)
[5] J. Kim, M. Zukic, M.W. McColgan, and D.G. Torr: (1993).
[6] J. I. Larruquert, A. M. Malvezzi, A. Giglia, J. A. Aznarez, L. Rodriguez-de Marcos, J. A. Mendez and G. Crescenzio: Journal of Optics 16(12), 125713(2014).
[7] F. Bridou, M. Cuniot-Ponsard, J. M. Desvignes, A. Gottwald, U. Kroth and M. Richter: Applied Physics A 102(3), 641-649(2011).
[8] T. Siefke, S. Kroker, K. Pfeiffer, O. Puffky, K. Dietrich, D. Franta and A. Tunnermann: Advanced Optical Materials 4(11), 1780-1786(2016).
[9] D. L. Windt, "IMD—Software for modeling the optical properties of multilayer films." Computers in physics 12(4), 360-370(1998).
[10] Comsol Multiphysics. Available online: http://www.comsol.com.

AUTHOR INDEX